Lecture Notes in Computer Science 10161

Commenced Publication in 1973
Founding and Former Series Editors:
Gerhard Goos, Juris Hartmanis, and Jan van Leeuwen

Editorial Board

More information about this series at http://www.springer.com/series/7409

Mauro Dragoni · María Poveda-Villalón
Ernesto Jimenez-Ruiz (Eds.)

OWL: Experiences and Directions – Reasoner Evaluation

13th International Workshop, OWLED 2016
and 5th International Workshop, ORE 2016
Bologna, Italy, November 20, 2016
Revised Selected Papers

Springer

Editors
Mauro Dragoni
Fondazione Bruno Kessler
Povo, Trento
Italy

Ernesto Jimenez-Ruiz
University of Oslo
Oslo
Norway

María Poveda-Villalón
Universitad Politecnica de Madrid
Madrid
Spain

ISSN 0302-9743   ISSN 1611-3349   (electronic)
Lecture Notes in Computer Science
ISBN 978-3-319-54626-1   ISBN 978-3-319-54627-8   (eBook)
DOI 10.1007/978-3-319-54627-8

Library of Congress Control Number: 2017932791

LNCS Sublibrary: SL3 – Information Systems and Applications, incl. Internet/Web, and HCI

Printed on acid-free paper

This Springer imprint is published by Springer Nature
The registered company is Springer International Publishing AG
The registered company address is: Gewerbestrasse 11, 6330 Cham, Switzerland

Preface

The OWL: Experiences and Directions Workshop series is an international forum for the OWL community, where practitioners in industry and academia, tool developers, and others interested in making use of OWL present research advances, real and potential applications, share experiences, and discuss requirements for language extensions/modifications. OWLED 2016 was the 13th edition of this workshop and was held on November 20 in Bologna, Italy, co-located with the 20th International Conference on Knowledge Engineering and Knowledge Management (EKAW 2016).

The technical program featured 11 presentations of accepted full (8) and short papers (3) and one invited talk:

– Maria Keet (University of Cape Town, South Africa): "Test-driven Development of Ontologies"

This year, for the first time, we joined efforts with the OWL reasoner evaluation workshop (ORE), which aims at gathering solutions and experiences for particular reasoning tasks on specific problems. There were 13 paper submissions to the workshop, which were reviewed by at least three Program Committee members. Reviews were aimed at constructive feedback and inclusiveness, in order to foster and strengthen the community spirit that characterizes OWLED. In all, 11 submissions were accepted to be presented during the workshop; three of them were categorized as *controversial* according to the reviewers' comments and were reserved a special slot within the workshop's program. We thank the Program Committee for their hard work in reviewing the submitted papers and for the useful feedback they gave to the authors. We would also like to thank the authors for submitting their papers and responding to the reviewers' comments in the final version. We further wish to thank the invited speaker for her inspiring talk. Our thanks also go to the University of Bologna, the local organizers of the 20th International Conference on Knowledge Engineering and Knowledge Management for helping us with the logistic organization of OWLED 2016, and the EKAW 2016 Organizing Committee. Finally, we would like to thank the development team of the EasyChair conference management system.

November 2016

Mauro Dragoni
María Poveda-Villalón
Ernesto Jimenez-Ruiz

Preface

The OWL: Experiences and Directions Workshop series is an international forum for the OWL community, where practitioners in industry and academia, tool developers, and others interested in making use of OWL present research advances, real and potential applications, share experiences, and discuss requirements for language extensions/modifications. OWLED 2016 was the 13th edition of this workshop and was held on November 20 in Bologna, Italy, co-located with the 20th International Conference on Knowledge Engineering and Knowledge Management (EKAW 2016).

The technical program featured 11 presentations of accepted full (8) and short papers (3) and one invited talk:

– Maria Keet (University of Cape Town, South Africa): "Test-driven Development of Ontologies"

This year, for the first time, we joined efforts with the OWL reasoner evaluation workshop (ORE), which aims at gathering solutions and experiences for particular reasoning tasks on specific problems. There were 13 paper submissions to the workshop, which were reviewed by at least three Program Committee members. Reviews were aimed at constructive feedback and inclusiveness, in order to foster and strengthen the community spirit that characterizes OWLED. In all, 11 submissions were accepted to be presented during the workshop; three of them were categorized as *controversial* according to the reviewers' comments and were reserved a special slot within the workshop's program. We thank the Program Committee for their hard work in reviewing the submitted papers and for the useful feedback they gave to the authors. We would also like to thank the authors for submitting their papers and responding to the reviewers' comments in the final version. We further wish to thank the invited speaker for her inspiring talk. Our thanks also go to the University of Bologna, the local organizers of the 20th International Conference on Knowledge Engineering and Knowledge Management for helping us with the logistic organization of OWLED 2016, and the EKAW 2016 Organizing Committee. Finally, we would like to thank the development team of the EasyChair conference management system.

November 2016

Mauro Dragoni
María Poveda-Villalón
Ernesto Jimenez-Ruiz

Organization

Executive Committee

General Chair

Mauro Dragoni   Fondazione Bruno Kessler, Italy

Program Chairs

María Poveda-Villalón   Universidad Politecnica de Madrid, Spain
Ernesto Jimenez-Ruiz   University of Oslo, Norway

OWLED Steering Committee

Melanie Courtot   BCCRC, Canada
Matthew Horridge   Stanford University, USA
Pavel Klinov   University of Ulm, Germany
Simon Jupp   EBI, UK
Mariano Rodriguez-Muro   IBM, USA
Bijan Parsia   University of Manchester, UK
Valentina Tamma   University of Liverpool, UK

Program Committee

Loris Bozzato   Fondazione Bruno Kessler, Italy
Francesco Corcoglioniti   Fondazione Bruno Kessler, Italy
Mauro Dragoni   Fondazione Bruno Kessler, Italy
Claudia D'Amato   University of Bari, Italy
Michel Dumontier   Stanford University, USA
Mariano Fernández López   Universidad San Pablo CEU, Spain
Daniel Garijo   Universidad Politecnica de Madrid, Spain
Rafael Gonçalves   Stanford University, USA
Pascal Hitzler   Wright State University, USA
Rinke Hoekstra   University of Amsterdam/VU University Amsterdam,
                                  The Netherlands
Ernesto Jimenez-Ruiz   Oxford University, UK
Yevgeny Kazakov   University of Ulm, Germany
C. Maria Keet   University of Cape Town, South Africa
Ilianna Kollia   National Technical University of Athens, Greece
Agnieszka Lawrynowicz   Poznan University of Technology, Poland
Despoina Magka   Facebook, UK
Francisco Martin-Recuerda   Universidad Politecnica of Madrid, Spain
Nicolas Matentzoglu   University of Manchester, UK

Contents

OntoJIT: Parsing Native OWL DL into Executable Ontologies in an Object Oriented Paradigm

Sohaila Baset[(⊠)] and Kilian Stoffel

Information Management Institute, University of Neuchatel, Neuchatel, Switzerland
{sohaila.baset,Kilian.Stoffel}@unine.ch

Abstract. Despite meriting the growing consensus between researchers and practitioners of ontology modeling, the Web Ontology Language OWL still has a modest presence in the communities of "traditional" web developers and software engineers. This resulted in hoarding the semantic web field in a rather small circle of people with a certain profile of expertise. In this paper we present OntoJIT, our novel approach toward a democratized semantic web where we bring OWL ontologies into the comfort-zone of end-application developers. We focus particularly on parsing OWL source files into executable ontologies in an object oriented programming paradigm. We finally demonstrate the dynamic code-base created as the result of parsing some reference OWL DL ontologies.

Keywords: Ontologies · OWL · Semantic web · Meta programming · Dynamic compilation

1 Background and Motivation

With a stack full of recognized standards and specifications, the Web Ontology Language OWL has made long strides to allocate itself a distinctive spot in the landscape of knowledge representation and semantic web. Obviously, OWL is not the only player in the scene; over the couple of last decades many other languages have also emerged in the ontology modeling paradigm. Most of these languages are logic-based formalisms with underlying constructs in first order logic [5,7,8,11] or in one of the description logic fragments like OWL itself [3,4] and its predecessor DAML+OIL [10]. Some frame-based languages have also seen some success in that area [12–14], in particular KL-One has integrated the automated deductive reasoning of logic-based languages into hierarchical semantic networks [9].

If we look at OWL characteristics; beside its strong expressive capabilities and logic based formalism, OWL has also got many flavors that are tailored to fulfill the different needs of ontology systems stakeholders [4]. These characteristics allowed OWL to stand out among its counterparts and OWL ontologies became dominant in a wide range of application domains. From the perspective of traditional software developers, however, these very same characteristics have

© Springer International Publishing AG 2017
M. Dragoni et al. (Eds.): OWLED-ORE 2016, LNCS 10161, pp. 1–14, 2017.
DOI: 10.1007/978-3-319-54627-8_1

contributed to a certain extent in augmenting the complexity surrounding OWL ontologies and logic-based formalisms in general.

We raise the issue of democratized semantics where a wider range of developers are invited to actively participate in the making process of semantic applications. Addressing this issue involves certainly more aspects than what we can cover in a single paper. In this paper, we rather start by whetting developers' appetite for ontologies by expressing them in a programming language –or paradigm– that developers are already comfortable with. For that purpose, we sketch our tool OntoJIT that parses existing OWL DL ontologies into executable fragments of code in C# while maintaining their semantics. We demonstrate the parsing results obtained and the limitations of the current state. We finally discuss some of the related projects and directions for future work.

2   Preliminaries

2.1   Executable Ontologies

Before being able to work on an ontology, inference engines require the ontology to be loaded into memory. This task is achieved by an ontology loader that transforms the ontology from its syntactic form e.g. RDF/XML into an in-memory representation. In the literature, there are two prominent in-memory representations for OWL ontologies: The first one is the abstract syntax tree AST model which is used in OWL API, previously known as the WonderWeb OWL API, [18,19]. The other representation model is the RDF triple and is the format that is adopted in Jena [15].

In our work, we look into the classification of in-memory ontology representations from a different perspective. More precisely we differentiate between two forms of in-memory representation: The passive form and the active form. To illustrate what we designate by each form, we consider the parsing output produced by Jena and OWL API; after the parsing step is completed, both parsing output models, i.e. AST or RDF graph, will eventually reside in the data segment of the program allocated memory waiting to be operated on by the inference engine and in that sense both are examples of the passive forms. In the active form, on the other hand, the output of the parsing step belongs to the code segment of the allocated memory. That is, the syntactic RDF/XML representation is transformed and loaded in memory as a set of executables.

Projecting the object oriented programming paradigm into this view of active in-memory ontology representation yields the term executable ontology that we first present in this paper. We can now think of OWL concepts and individuals as OOP classes and instances spread over code namespaces that can be compiled and run.

2.2   Meta Programming in Strongly Typed Languages

Parsing RDF/XML into executable ontologies clearly adds another layer of complexity into the already non-trivial parsing task. It requires dynamically generating code statements that are equivalent to the RDF triple being parsed. In such settings, the deployment of meta programming techniques proves advantageous. Meta programming refers to the programming paradigms and the means by which a program has knowledge of itself or can manipulate itself. To that end, meta programs are programs that write programs. Examples of meta programs are optimizers, partial evaluation systems and program transformers [20]. There exists many classification of meta programs, among them is the static vs run-time classification i.e. whether the produced output program is written to disk or dynamically compiled at run-time, the manually vs automatically annotated classification i.e. whether the staging annotations are placed directly by the programmer or produced by an automatic process and finally the homogeneous vs heterogeneous programs which concerns whether or not the meta language is the same as the program output language [20]. Our proposed OntoJIT RDF/XML parser is a manually annotated, run-time heterogeneous meta program.

Like many paradigms in software development, Meta programming is an approach that is not equally supported by all programming languages. Some languages, such as CaML [24], are designed with meta programming in the core of their philosophy. Dynamic languages like Prolog and smalltalk have fundamental meta programming features [21]. Macros in Lisp and Scala also provide strong support for meta programming [22,23], whereas Python programmers usually use meta classes. When it comes to strongly typed languages, however, the emphasis on meta programming features becomes less evident. This does not mean that meta programming is not supported in many of these languages; C++ offers templates for meta programming [31], Java programs have annotations [30] and .Net languages use annotations and/or reflection to produce meta programs [32]. Indeed, the parser presented here was realized using one of the meta programming libraries offered in .Net [33].

3   OntoJIT Parser

Parsing OWL source files into executable source code is the first step of an ongoing effort to bring ontologies into the table of application developers. The overall goal of this effort is not limited to parsing ontologies into compiled source code, the real interesting part is the potential reasoning possibilities over this newly created eco-system of executable ontologies; hence the name OntoJIT refers to just in time ontologies and is inspired from the dynamic "Just in Time" compilation in .Net languages. The OntoJIT parser we present here is written C#. It produces compiled source code in the form of dynamic linking libraries or executables and it can also produce C# source files as an intermediate output. In the following sections we discuss some of the key points in the design and implementation of OntoJIT parser.

3.1   Parsing OWL Files

OWL Graph Traversal. Most existing OWL parsing tools use a recursive depth first search to perform a one-pass traversal of OWL source. This seems like an elegant approach for a streaming-like parsing; the DFS serves as a serialization technique and for each construct visited in the source, a corresponding node or edge is attached to the OWL graph being constructed in memory. However, when parsing output is an executable, the pure DFS approach is unfortunately insufficient. Deciding on the corresponding code statements to a syntactic construct requires all information related to this construct to be available at node processing time; which is clearly not the case with the inter-node associativity present in OWL source documents. Here we have two approaches to overcome this limitation, first approach is to use multiple-pass traversal to guarantee that we have complete information before generating the corresponding output. This approach is clearly less efficient compared to the one-pass traversal both in execution time as well as in space complexity since it requires maintaining the intermediate state of nodes being parsed over many passes. The other approach, which is the one used in this paper, is to combine pre-order DFS traversal with look up operations when necessary. The parser presented here is built to read RDF/XML syntax; in that case, the look-up operations are simply forward jumps within to the RDF/XML child nodes and the set of possible look-ups is limited assuming prior knowledge of the associations patterns of OWL nodes.

Import Closure. Ontology modeling practices share some of the design principles with software engineering, mostly with regards to the re-usability of existing ontologies. An ontology is not isolated from other ontologies, it builds up on top of other already existing ones. In ordinary programming languages, this corresponds into importing packages or libraries and in OWL, to using import keyword to allow the usage of terms defined in the imported namespaces. Keeping on with this analogy, the OntoJIT parser treats imported namespaces in OWL source as namespaces in the target output code. When the parser reads an `owl:imports` term, it triggers a recursive call to the main parsing routine for all imported ontologies until an import closure is achieved.

3.2   OWL to OOP Mapping

When comparing the expressiveness aspect of OWL to that of formal programming languages, programming languages rank way below than even the most restricted profile of OWL. The semantic richness of OWL DL ontologies makes it difficult to find an OOP counterpart for each OWL DL semantic construct. Furthermore, there are some fundamental differences between the two schools of modeling such as the notion of disjoint classes, inheritance model and many others. When mapping OWL DL to OOP, our goal was to exploit the native programming language constructs while at the same time trying not to violate the OOP design principles. Although the mapping seems self-evident in some parts e.g. `owl:class` as an OOP class, `rdfs:subClassOf` as an OOP class inheritance relation and OWL individuals as instances of OOP classes; finding the right mapping becomes more problematic when we consider OWL DL terms such as: `owl:disjointWith`, `owl:sameAs` and `owl:equivalentClass`. One could still create native constructs that are semantically equivalent to such terms by enforcing

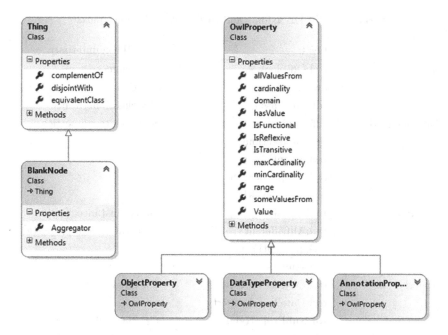

Fig. 1. The initial output scheme in OntoJIT.

some design patterns and constraints but this approach has some consequences that we will discuss in one of the following sections. One other possibility is to rely on annotations to express all OWL terms that are foreign in the OOP language, but in plain OOP terms this means that most of the modeled information about an object is laid outside of it and is not directly accessible via its properties. Instead, in OntoJIT parser, for the major part of OWL terms, meta properties are created that form the bases for mapping OWL concepts, properties and restrictions. The meta properties are defined in the top hierarchy level and are then inherited by all parsed classes afterwards and masked where necessary. One important thing to clarify is that the term "meta" used here refers to a completely different sense than the programming technique discussed earlier, the usage of the term here is rather functional; the idea is that these meta properties would cover up for the missing explicit semantics in the formal language constructs, and the full interpretation of the meta properties semantics is to be realized by an inference component on top of the parsing layer. Figure 1 shows the initial output scheme in OntoJIT where meta properties are first defined.

Blank Nodes. Just like in RDF/XML, OntoJIT uses blank nodes to express a property restriction or class description axioms. Though in our implementation, blank nodes are not anonymous; they are created as class definitions with automatically (and deterministically) generated names to make them available for subsequent inference tasks. On the other hand, since these nodes are not

Table 1. OWL DL axioms and their OntoJIT counterparts

Axiom	OWL	OntoJIT counterpart
Ontology	owl:Ontology	Code namespace
Class	owl:class	C# class
	rdfs:subclass	C# class inheritance
Class description	rdfs:equivalentClass	Static meta properties
	owl:intersectionOf	
	owl:unionOf	
	owl:complementOf	
	owl:disjointWith	
Individual	Individual	Object instance
	owl:AllDifferent	Non-static meta properties
	owl:differentFrom	
	owl:sameAs	
Property	owl:ObjectProperty	C# class
	owl:DataTypeProperty	
	rdfs:subPropertyOf	C# class inheritance
Property association	rdfs:range	Static meta properties
	rdfs:domain	
Property restriction	rdfs:cardinality	Static meta properties
	rdfs:hasValue	
	rdfs:someValuesFrom	
	rdfs:allValuesFrom	
Property Description	owl:FunctionalProperty	Static meta properties
	owl:InverseFunctionalProperty	
	owl:SymmetricProperty	
	owl:TransitiveProperty	
Property relations	owl:inverseOf	Static meta properties
	owl:subPropertyOf	
	owl:equivalentProperty	

explicitly part of the ontology class definitions, these classes get the private access modifier and are therefore invisible from outside the namespace they belong to.

Semantic Equivalence. The semantic expressiveness of the source ontology is preserved with the aid of meta properties. As stated earlier, the role of meta properties is to cover up for missing explicit semantics in the formal language constructs, i.e., when there is no programming language counterpart for an axiom in the source ontology or when relying on the programming language to express an axioms would interfere with the Open World Assumption OWA. For example, the property association axiom `rdfs:range` could be easily parsed into the data type of the property in the class definition where it belongs to. While this is the norm from a strict modeling perspective, it does not conform to OWA inference mechanism. According to OWA, having two different fillers for the range property is perfectly fine as long as they are not stated to be distinct; whereas this would certainly not pass type checking performed by an OOP language compiler.

It is also worth mentioning that OntoJIT, in its current state, supports OWL $\mathcal{SHOIN}(\mathcal{D})$ DL profile. Parsing ontologies with OWL 2 DL $\mathcal{SROIQ}(\mathcal{D})$ extensions [6], like for example General Concept Inclusion axioms, has not been tested. Table 1 lists OWL DL axioms and their OntoJIT C# counterparts.

4 Demonstrations

To test the parsing process introduced in the previous section, we used the two famous OWL DL Pizza[1] and wine[2] ontologies. These ontologies are relatively small in size but they are pretty expressive as they were created for the purpose of demonstrating the different capabilities of OWL DL and they would therefore be helpful in validating the parsing routine.

Formally proving the semantic equivalence of an OWL DL ontology and the corresponding executable produced by OntoJIT would require at least comparing results of some inference tasks over the two formats which, at this stage of our work, is not possible yet. Instead in this section we demonstrate some code snippets examples of the parsing results and their OWL counterparts.

```
/// <summary>
/// Any Pizza that is not a VegetarianPizza
/// </summary>
public class NonVegetarianPizza : Thing
{
    public NonVegetarianPizza(){}

    public static object equivalentClass
    {
        get{ return "Blank23";}
    }
    public static object disjointWith
    {
        get{return "VegetarianPizza";}
    }
}
```

Fig. 2. Non-vegetarian pizza class definitions (a).

OWL Classes. To start with, we consider the example of non-vegetarian pizza definition in the pizza ontology. The produced code snippet is demonstrated in Figs. 2 and 4 and the original OWL source is shown in Fig. 3. The following is the DL notation of the same information:

$$NonVegetarianPizza \equiv \neg VegetarianPizza \sqcap Pizza$$

$$NonVegetarianPizza \sqcap VegetarianPizza \equiv \bot$$

---

[1] www.protege.stanford.edu/ontologies/pizza/pizza.owl.
[2] www.w3.org/TR/owl-guide/wine.rdf.

In the "NonVegetarian" class definition in Fig. 2. We can see that the `owl:equivalentClass` term is expressed by mean of the meta property equivalentClass which returns as object (of the RDF triple) a blank node identifier "Blank23". The "Blank23" stands for the anonymous class representing $\neg VegetarianPizza \sqcap Pizza$ that in turn is defined as the intersection of another blank node "Blank22" with the class pizza. Finally "Blank22" is defined as a blank node class with the "ComplementOf" and "VegetarianPizza" meta properties values. As mentioned earlier, the meta properties used in expressing the definitions are essential for substituting for the explicit semantics that are not available as native language constructs. The examples shown here use a textual representation of the values for these properties, in fact these values are just the handles to the created types in the code namespace and are available for later use by the inference component in runtime via reflection.

```
Class: NonVegetarianPizza

     Annotations:
          rdfs:label "PizzaNaoVegetariana"@pt,
          rdfs:comment "Any Pizza that is not a VegetarianPizza"@en

     EquivalentTo:
          Pizza
          and (not (VegetarianPizza))

     DisjointWith:
          VegetarianPizza
```

Fig. 3. Non-vegetarian pizza description in manchester syntax

```
private class Blank23 : BlankNode
{
    public static object Aggregator
    {
        get
        {
            return OwlVocabulary.intersectionOf;
        }
    }

    public static object CollectionItems
    {
        get
        {
            return "Blank22;Pizza;";
        }
    }
}
```

```
private class Blank22 : BlankNode
{
    public static object Aggregator
    {
        get
        {
            return OwlVocabulary.ComplementOf;
        }
    }

    public static object CollectionItems
    {
        get
        {
            return "VegetarianPizza;";
        }
    }
}
```

Fig. 4. Non-vegetarian pizza class definitions (b).

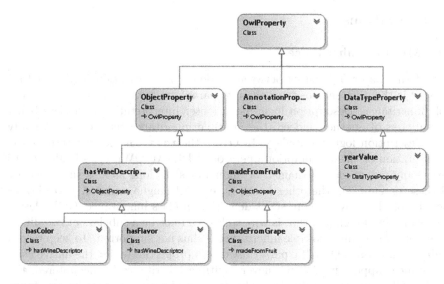

Fig. 5. Reduced sketch of the hierarchy of wine ontology parsed properties

OWL Properties. Just like classes, OWL properties have hierarchical structures. They also have characteristics such as domain, range and cardinality. This is well reflected into OntoJIT executable ontologies. The parser starts with initial hierarchy shown in Fig. 1 and expands it as the parsing continues. Parsed classes would then have instances of these properties to express relation between individuals. Figure 5 shows a reduced (incomplete) sketch of the properties hierarchy in the wine ontology. The code snippets in Fig. 6 show the definitions for some properties along with their characteristics. The characteristics of a property are supposed to be shared among all its instances and are therefore declared static whereas the instance value of the property is a non-static variable.

```
public class hasFlavor : hasWineDescriptor
{
    public hasFlavor() {}
    public static object FunctionalProperty
    {
        get { return true; }
    }
    public static object range
    {
        get { return "WineFlavor"; }
    }
}
```

```
public class hasWineDescriptor : ObjectProperty
{
    public hasWineDescriptor(){}
    public static object domain
    {
        get {return "Wine";}
    }
    public static object range
    {
        get{return "WineDescriptor";}
    }
}
```

Fig. 6. OntoJIT property classes for hasFlavor property and its parent property.

5   Limitations

5.1   Multiple Inheritance

One of the major differences between modeling in description logic and that in OOP is the different positions the two paradigms have with regards to multiple inheritance. Description logic has a looser interpretation of a class being the subclass of another; indeed, the multiple inheritance term does not really fit in description logic vocabulary. In OWL, the `rdfs:subclassOf` term is the manifestation of the subsumption operator of DL. An OWL class is allowed to have many parent classes (named or anonymous) as long as it is subsumed by all these parents. On the other hand, pure OOP languages like C# or Java – though not all – have a more strict definition of class inheritance, OOP classes are disjoint by design and that is why a class can not be a subclass of two different parent classes and multiple inheritance is thus not supported. To keep record of all parent classes, OntoJIT parser uses meta properties beside the native class inheritance support in C#, whenever multiple inheritance is encountered, the subClassOf property is extended. This workaround suffers from inconsistency but is still preferable over relying on interfaces where one could use interface declarations instead of classes to reflect OWL hierarchies. The problem with the interface approach is that interfaces are abstract and thus are not instantiable and one would need to create a shadow class for each declared interface. This can quickly become an overkill and unscalable when considering relatively complex ontologies with a lot of blank nodes. Left with these two not really optimal solutions, the pursuit of a more elegant one is still an open question.

5.2   Import Closure

The approach taken to handle the `owl:imports` terms is a little bit a minimalist approach for one reason; it doesn't handle the case where the ontology being parsed is an OWL DL or OWL light ontology and the imported ontology is an OWL Full one. The parser presented here is mainly concerned with OWL DL or Light profiles and more investigations and analysis are necessary before attempting on parsing an OWL FULL ontology. In this case, the parser is not able to process OWL Full constructs and will therefore skip them. This for sure would have an impact on the soundness of the reasoning results but as reasoning is not yet in the scope of the current state of OntoJIT, this is something to be addressed again as the work in the project advances.

6   Related Work

The difficulty of utilizing OWL ontologies in conventional software projects was behind the work presented in [29]: The authors demonstrate some of the fundamental differences between the "subject-predicate-object" school of modeling (with persistent triple-stores) and the object oriented school (with normalized relational databases). According to the authors, the combined use of ontologies with standard programming practices would enable the development of semantic-rich enterprise applications and they suggest a framework for translating some ontology constructs into Enterprise Java Beans. In [28], the primary intention is to provide guidance on how to build real-world semantic web applications. The authors draw analogy between deploying ontologies as high-level models in software development and the approach used in Model Driven Architecture MDA. They also suggest a software architecture for web services and agents for the semantic web driven by domain ontologies. [26] proposes a hybrid modeling software framework that combines the object oriented representation of a domain with its ontological representation. The authors analyze the advantages and disadvantages of such hybrid modeling approach by means of a case study of a large medical records system. There exist as well many API projects to integrate OWL ontologies into application development. The OpenRDF API[3] along with its satellite projects Elmo/Alibaba[4], provides object triples mapping for creation of flexible RDF-based applications. Another object-oriented API for managing RDF is ActiveRDF [25], it offers schema-free manipulation and querying of RDF data while conforming to RDF(S) semantics. Overall, OWL to UML mapping has a good share of papers in the literature. In [27] A UML-based visualization of OWL DL ontologies is presented. The work done in [34] provides a rigorous comparison between UML and OWL as two flagship languages for artificial intelligence and software engineering communities; the authors argue that based on the core definitions of ontologies and models, none of the common informal distinctions made between the two terms is actually justifiable. Instead, ontologies themselves are to be regarded as models. Further more, without changes to the currently used ways of distinguishing between models and ontologies the confusion around the two terms will continue to arise.

On the technical side, one particular project that addressed the idea of mapping OWL ontologies into JAVA OOP classes is in [16]. The main aim of the project was aiding semantic application development and the approach taken was to try to stretch the expressiveness of modeling in Java to that of OWL DL by enforcing some constraints and design patterns: Interfaces for multiple inheritance, special listeners on property accessors, type checking for domain and range properties, etc. While we see the motivation behind this approach, we believe that it entails some twisting in the interpretation of OO design principles and what is originally supposed to be explicit semantics in OWL is becoming rather implicit and dependent on the interpretation of the "special purposes" patterns used. Another observation is that this approach would work just fine as long as only the modeling part is concerned but if performing inference tasks is part of the deal, then more caution is necessary. Relying merely on native Java constructs to translate OWL DL means in a certain way delegating the responsibility of enforcing restrictions and properties characteristics to the compiler, which is not exactly the point of properties and restrictions axioms from an

---

[3] OpenRDF, http://www.openrdf.org/.
[4] https://bitbucket.org/openrdf/alibaba.

open world reasoning perspective. Another related project is in [17]. The authors proposed an initial Python metaclass-based representation of OWL ontologies that offer class declaration and instance creation. Their prototype also allows integrating an OWL DL reasoner with their metaclass representation.

7   Conclusion and Future Work

In this paper we presented a novel approach into democratized semantics by bringing OWL ontologies into the context of programming languages. We also reported on our experience in automatically parsing ontologies into executables. Since the project is in its early stage, there is a lot on the road map for Onto-JIT; mainly exploring the reasoning possibilities over executable ontologies and potential advantages or drawbacks this can bring. One idea here is that with run-time dynamic compilation of modern programming languages, the generated source code can change and adapt at run time. In that sense, executing the ontologies would result in spanning the source code as more explicit information are inferred from initial implicit semantics. Another interesting possibility is to exploit hierarchical self-organizing models when inferring class hierarchy using meta properties as features of asserted input class definitions.

Apart from reasoning, there is also more to investigate on the subject of the chosen programming paradigm; the OOP paradigm was a close fit from the modeling perspective but applying the same idea in a declarative paradigm would also be of interest.

In the long run, we believe that even though the presented idea of democratized semantics is in its infancy stage: the more research we do in this direction the more potentials arise in the two universes of application development and knowledge representations alike.

References

1. Horrocks, I., Patel-Schneider, P.F., van Harmelen, F.: From SHIQ and RDF to OWL: the making of a web ontology language. Web Semant. Sci. Serv. Agents World Wide Web 1(1), 7–26 (2003). ISSN 1570–8268
2. Baader, F., McGuinness, D., Nardi, D., Patel-Schneider, P.: The Description Logic: Handbook Theory, Implementation and Applications. Cambridge University Press, Cambridge (2002)
3. McGuinness, D.L., Van Harmelen, F.: OWL web ontology language overview. W3C Recommendation 10(10), 2004 (2004)
4. OWL 2 Profiles, OWL 2 Web Ontology Language Profiles, W3C Recommendation (2009)
5. Delugach, H.: ISO/IEC WD 24707 Information technology Common Logic (CL) A Framework for a Family of Logic-Based Languages. Pacific Northwest National Laboratory, Chantilly, VA 7 (2004)
6. Horrocks, I., Kutz, O., Sattler, U.: The even more irresistible SROIQ. Kr 6, 57–67 (2006)

7. Lenat, D.B., Guha, R.V.: The evolution of CycL, the Cyc representation language. ACM SIGART Bull. **2**(3), 84–87 (1991)

8. Genesereth, M.R., Fikes, R.E.: Knowledge interchange format-version 3.0: reference manual (1992)

9. Brachman, R.J., Schmolze, J.G.: An overview of the KL-ONE knowledge representation system. Cognit. Sci. **9**(2), 171–216 (1985)

10. Horrocks, I.: DAML+OIL: a description logic for the semantic web. IEEE Data Eng. Bull. **25**(1), 4–9 (2002)

11. Niles, I., Pease, A.: Towards a standard upper ontology. In: Proceedings of the International Conference on Formal Ontology in Information Systems, vol. 2001. ACM (2001)

12. Kifer, M., Lausen, G., James, W.: Logical foundations of object-oriented and frame-based languages. J. ACM **42**, 741–843 (1995)

13. Chaudhri, V.K., Farquhar, A., Fikes, R., Karp, P.D., Rice, J.P.: OKBC: a Programmatic foundation for knowledge base interoperability. In: AAAI 1998 Proceedings (1998)

14. Clark, P., Porter, B., Works, B.P.: KM-the knowledge machine 2.0: users manual, vol. 2, p. 5. Department of Computer Science, University of Texas at Austin (2004)

15. Carroll, J.J., et al.: Jena: implementing the semantic web recommendations. In: Proceedings of the 13th International World Wide Web Conference on Alternate Track Papers & Posters. ACM (2004)

16. Kalyanpur, A., et al.: Automatic mapping of OWL ontologies into Java. In: SEKE, vol. 4 (2004)

17. Babik, M., Hluchy, L.: Deep integration of python with web ontology language. In: Proceedings of the 2nd Workshop on Scripting for the Semantic Web (2006)

18. Bechhofer, S., Carroll, J.J.: OWL DL: trees or triples? In: Proceedings of the Thirteenth International World Wide Web Conference (WWW 2004) (2004)

19. Bechhofer, S., Volz, R., Lord, P.: Cooking the semantic web with the OWL API. In: Fensel, D., Sycara, K., Mylopoulos, J. (eds.) ISWC 2003. LNCS, vol. 2870, pp. 659–675. Springer, Heidelberg (2003). doi:10.1007/978-3-540-39718-2_42

20. Sheard, T.: Accomplishments and research challenges in meta-programming. In: Taha, W. (ed.) SAIG 2001. LNCS, vol. 2196, pp. 2–44. Springer, Heidelberg (2001). doi:10.1007/3-540-44806-3_2

21. Abramson, H., Rogers, M.H.: Meta-Programming in Logic Programming. MIT Press, Cambridge (1989)

22. Burmako, E.: Scala macros: let our powers combine!: on how rich syntax and static types work with metaprogramming. In: Proceedings of the 4th Workshop on Scala. ACM (2013)

23. Hoyte, D.: Let Over Lambda. Lulu.com (2008)

24. Pottier, F.: An overview of CML. Electron. Notes Theor. Comput. Sci. **148**(2), 27–52 (2006)

25. Oren, E., Delbru, R., Gerke, S., Haller, A., Decker, S.: Object-oriented semantic web programming. In: Proceedings of the 16th International Conference on World Wide Web (WWW 2007), pp. 817–824. ACM, New York (2007). doi:http://dx.doi.org/10.1145/1242572.1242682

26. Puleston, C., Parsia, B., Cunningham, J., Rector, A.: Integrating object-oriented and ontological representations: a case study in Java and OWL. In: Sheth, A., Staab, S., Dean, M., Paolucci, M., Maynard, D., Finin, T., Thirunarayan, K. (eds.) ISWC 2008. LNCS, vol. 5318, pp. 130–145. Springer, Heidelberg (2008). doi:10.1007/978-3-540-88564-1_9

27. Brockmans, S., Volz, R., Eberhart, A., Löffler, P.: Visual modeling of OWL DL ontologies using UML. In: McIlraith, S.A., Plexousakis, D., Harmelen, F. (eds.) ISWC 2004. LNCS, vol. 3298, pp. 198–213. Springer, Heidelberg (2004). doi:10. 1007/978-3-540-30475-3_15
28. Knublauch, H.: Ontology-driven software development in the context of the semantic web: an example scenario with Protege/OWL. In: 1st International Workshop on the Model-Driven Semantic Web (MDSW2004), Monterey, California, USA [WWW document] (2004). http://www.knublauch.com/publications/MDSW2004. pdf
29. Athanasiadis, I.N., Villa, F., Rizzoli, A.-E.: Ontologies, JavaBeans and relational databases for enabling semantic programming. In: 31st Annual International Computer Software and Applications Conference (COMPSAC 2007), vol. 2. IEEE (2007)
30. Czarnecki, K., Eisenecker, U.W.: Generative Programming: Methods, Tools and Applications. Addison-Wesley, Reading (2000). Edited by Goos, G., Hartmanis, J., van Leeuwen, J
31. Abrahams, D., Gurtovoy, A.: C++ Template Metaprogramming: Concepts, Tools, and Techniques from Boost and Beyond. Pearson Education, Stoughton (2004)
32. Schult, W., Polze, A.: Aspect-oriented programming with C# and .net. In: Proceedings of the Fifth IEEE International Symposium on Object-Oriented Real-Time Distributed Computing (ISORC 2002). IEEE (2002)
33. Ganz, Carl., Jr.: Runtime code compilation. In: Pro Dynamic. NET 4.0 Applications, pp. 59–75. Apress (2010)
34. Atkinson, C., Gutheil, M., Kiko, K.: On the relationship of ontologies and models. WoMM **96**, 47–60 (2006)

Healthy Lifestyle Support:
The **PerKApp** Ontology

Tania Bailoni, Mauro Dragoni[(✉)], Claudio Eccher, Marco Guerini,
and Rosa Maimone

FBK-IRST, Trento, Italy
{tbailoni,dragoni,eccher,guerini,rmaimone}@fbk.eu

Abstract. Healthy lifestyle is not only a today trend fostered by the explosion of gluten-free foods (or similar) or by the presence on the market of many devices for monitoring how many steps you do during a day and how many calories you spent in the last twenty-four hours. Following a healthy lifestyle means also to prevent diseases as consequence of an incorrect diet or to avoid chronic pathologies that may occur after sensitive surgeries. In this paper, we present the first version of the **PerKApp** ontology. Here, we model concepts representing detailed foods properties, with the goal of supporting the construction of intelligent interfaces for domain experts. This ontology is part of the **PerKApp** project aiming to provide a full-fledged platform supporting the remote lifestyle monitoring of users by providing real-time feedback through persuasive context-based messages when necessary. Beside the ontology, the paper will also provide an overview of the **PerKApp** project and how the presented ontology will be used.

1 Introduction

Diets and physical activity play a crucial role for a long and healthy life. Best practices are available in guidelines and expert recommendations regarding healthy lifestyle that people should adopt for maintaining their physical and mental well being. This way, they will be able to prevent cognitive decline, obesity, disability, and death from major chronic diseases: diabetes, cardiovascular disease, and several forms of cancer, just to mention a few. However, engaging people in developing and maintaining healthier patterns of living is a challenging task.

ICT-based persuasion systems can be effective tools to persuade and motivate people to change their behavior. Such systems are able to collect and reason on user's data gathered from personal devices, off-the-shelf wearable sensors, and external sources (e.g., electronic health-care records). By exploiting these data, persuasive systems can generate effective personalized recommendations by adapting the message generator in response to the modification of the environment and the user status. To carry out this task, a persuasion tool must rely on a considerable amount of knowledge from different domains (e.g. user attitudes, preferences and environmental conditions, etc.) for suggesting the

M. Dragoni et al. (Eds.): OWLED-ORE 2016, LNCS 10161, pp. 15–23, 2017.
DOI: 10.1007/978-3-319-54627-8_2

behavior to adopt and for justifying such suggestions. Examples are food content and nutrients, physical activities accompanied by information concerning their categorization and effort, user attitudes and preferences, linguistic knowledge, and smart environment information (places, weather, etc.). As we may notice, these systems can greatly benefit from the adoption of an ontological approach to model knowledge, ensuring disambiguation of terms and formal definition of concepts and relations of the domains of discourse, which the system can exploit for reasoning purposes.

In this paper, we present the first version of the PerKApp ontology aiming to describe food properties and to support the construction of intelligent interfaces allowing domain experts to model monitoring rules for recommending healthier life styles. The ontology is part of the several knowledge bases modeled in the PerKApp project [1], which aims to provide a full-fledged platform supporting the monitoring of citizens and patients lifestyles and the provision of real-time feedback through persuasive context-based messages when the need for an intervention is detected.

Section 2 provides a brief overview of the main ontologies concerning the food domain. In Sect. 3, we present the PerKApp project. Then, in Sect. 4 the PerKApp ontology is described, while, in Sect. 5, we show how the ontology is used within the PerKApp architecture. Finally, Sect. 6 concludes the paper.

2   Related Work

Literature about food ontologies is not new and some works already provided useful artifacts. In this section, we briefly resume the most relevant work in this direction.

In [2] the authors describe food intake patterns identified by applying new food categories. New food groups were formed using a systematic approach involving the consideration of (i) nutrient composition and energy density, (ii) current scientific evidence of health benefits, and (iii) culinary use of each food. In this way the researchers identified 17 food groups.

Regarding the use of ontologies, in [3] it is presented a process for a rapid prototyping of a food ontology oriented to the nutritional and health care domain that is used to share existing knowledge. The aim of this ontology is to present a complete description of food with nutritional information, type, nutrients, and the recommended daily or weekly quantity to be consumed in a healthy diet for people with diabetes. The main steps of this process consist in: (i) identifying the domain and its rules; (ii) finding nouns used in common language for generic food; (iii) defining relations among food and properties and, also, among different foods (i.e. it does not exist any difference between two apples). The described ontology contains 177 classes, 53 properties of foods, and 632 relations.

The contribution presented in [4] discusses the design and development of a food-oriented ontology-driven system (FOODS), used for food or menu planning in a restaurant, clinic/hospital, or at home. FOODS comprises (i) a food ontology, (ii) an expert system using such an ontology and some knowledge about cooking methods and prices, and (iii) a user interface suitable for users with different levels of expertise.

Other works use ontologies for delivering personalized and customized information. The work presented in [5] focuses on the integration of different domain ontologies, like food, health, and nutrition, in order to help personalized information systems to retrieve food and health recommendations based on the user's health conditions and food preferences.

Finally, in [6] there are described the design steps, the working mechanism, and the case of use of the Ontology-Driven Mobile Safe Food Consumption System (FoodWiki) using semantic matching. The system is designed to evaluate commercial packaged food products and suggesting the selected product's appropriateness to food consumers according to their health conditions or intolerance. The Food Ontology Knowledge Base (FOKB) is also presented: it contains four main classes, 58 sub-classes, 15 object type properties and 17 sub-object type properties, 12 data type properties, 1530 individuals with annotation type properties, and 210 semantic rules.

The principal novelty of the presented ontology consists in providing a knowledge schema that is able not only to describe detailed information about foods, but also to support reasoning activities on users' behaviors and the creation of smart interfaces for creating rules for monitoring users.

3   The PerKApp Project

The PerKApp project[1] aims to merge the advantages of using diverse knowledge representation and reasoning techniques with rich persuasive natural language generation approaches. It is composed of two parts:

1 An application for personal devices, able to collect data from the user (e.g., food intake), wearable sensors (e.g., fitness trackers), and contextual data (e.g., the weather, the proximity of fitness facilities) and to notify persuasive and motivational messages by exploiting different representation mechanisms: textual, speech, video, and graphical alerts.
2 A core persuasive component that combines data and knowledge to generate effective persuasive messages customized to the user needs, attitudes and preferences, conveyed through multiple communication channels and modalities, dynamically selected exploiting contextual information (e.g., user's location and activity).

Three main concepts drive the message generation process:

external event: an event that occurs in the real world acting as trigger for the system (e.g., a timer, a user eating too much food or performing too low in physical activity).
communicative goal: the top most intention of the system that drives its planning (e.g., diverting from an actual unhealthy behavior).

---

[1] https://perkapp.fbk.eu.

persuasive goal: goals within the persuasion engine activated according to the top communicative goal, and representing partial "plans" to fulfill it.

The persuasion engine combines different knowledge bases for inferring the right content, type, and timing of the messages sent to users. In particular, the full-set of the exploited knowledge contains:

- "static" domain knowledge: i.e., the knowledge describing the objects of the domain of interest and their relationships, such as the food ontology presented here.
- The "dynamic" user model, i.e., the knowledge about users that may change over time as a consequence of external events (e.g., the health status) or system's actions (e.g., after the adoption of suggested behaviors).
- "environment" information concerning the context and the environment around users: city maps, information about sport facilities, areas for physical activities, etc.
- The "linguistic" model: i.e., the knowledge about the linguistic terms (noun, verbs, adjectives, etc.) and phrases more adapt for the domain of intervention, augmented with information about evocative qualities of such terms (e.g., sentiment bearing words, adjectives that represent level of intensity in physical activity).

4   The PerKApp Ontology

The development of the PerKApp ontology followed the need of providing a knowledge artifact able not only to provide a representation of domains concerning healthier lifestyle, but, also, to support further activities like, for example, remote medical monitoring. As we discussed in Sect. 2, ontologies available in domains connected with wellness and health lifestyle have been designed with different aims. For example, ontologies concerning sport activities are created with a focus on classifying data without connecting each activity with potential problems or benefits associated with the health status of a user. Similarly happened for ontologies concerning foods.

In the PerKApp project, we decided to model an ontology with a focus on the connection between diet and physical behavior with people health. The development of our ontology has been driven by the following main questions:

- Which information are needed for having a detailed description of each food?
- Which concepts are necessary for supporting the design of rules allowing user monitoring?
- Which data have to be provided by users for allowing reasoning tasks?

At this stage, the ontology has not been connected with fundamental ontologies like DOLCE [7]. The reason is that this first version of the ontology will be extended for covering further domains that will be of interest for the PerKApp project. Thus, the alignment task has been postpone as future work.

Figure 1 shows a general overview of the ontology with the main concepts.

Below, we distinguished concepts in three main categories and we provided the semantic meaning of the most important entities.

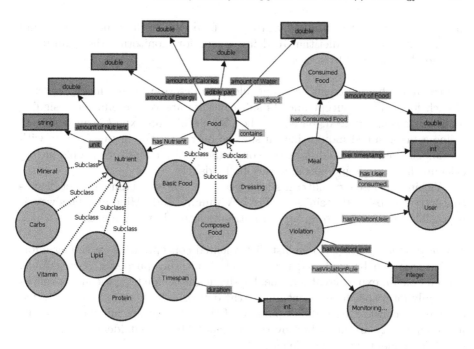

Fig. 1. Overview of the PerKApp ontology.

4.1   Diet-Oriented Concepts

These concepts are used for classifying foods and for describing, in detail, their composition. According to project aims, modeling only the category of each food and its calories is not enough. Thus, what we need is to model all possible information connected with food compositions. The main concepts belonging to this category are the following.

"Food". Trivially, this is the root concept of all foods contained into the ontology. As specifications of the concept "Food", we defined three main sub-concepts: "Basic Food", "Composed Food", and "Dressing". With "Basic Food", we intended to model foods that are described within the resources we adopted for creating the ontology. For all entities of type "Basic Food", we are able to provide a full description of their properties, as described later. While, the "Composed Food" concept represents entities, like dishes, that are composed by two or more concepts of type "Basic Food". The "Dressing" concept is exploited for modeling different dressings and toppings commonly available. This type of concepts has been thought for easing user monitoring activities. This distinction allows to map common-sense dishes that can be provided by users, as "Pasta alla Carbonara", to the set of basic foods accompanied with nutrition information.

Concerning the "Basic Food" concept, many specifications have been defined. The reader may consult them into the ontology. Here, we only want to mention that many of the modeled categories have been created for facilitating the user monitoring task. For example, by grouping foods under the concept "Fats and Oils", it is easy for a dietitian to define a rule for monitoring the quantity of fatty foods consumed by a user.

"Nutrient". The "Nutrient" concept allows to describe fine-grained properties of each food. This concept and its sub concepts are very useful for designing monitoring rules. Indeed, based on the user profile, an expert may decide to define a fine-grained monitoring on specific nutrients. For instance, users with cardiological issues should limit the consumption of fat foods, or people having calcium shortage should consume specific foods, etc.

Within our ontology, nutrients have been then classified among a set of sub-concepts: "Carbs", "Protein", "Lipids", "Vitamin", and "Mineral". This way, it is possible to monitor also groups of nutrients instead of single ones.

"Timespan" and "Meal". The last diet-oriented concepts concern timing information that can be exploited for different monitoring purposes. For example, when a domain expert creates a rule, he/she might decide to mark such a rule as valid only for a specific moment during the day, or as a check for a given period of time. By considering the diet purpose of the ontology, specific moments during a single day are described by the concept "Meal" that identifies the most common moments when people eat.

Instead, concerning specific time period, we defined the concept "Timespan". Such a concept is instantiated with individuals defined through the interface provided to domain experts. The "Timespan" concept is in relation with the datatype property "duration" allowing to specify the number of seconds for which a specific timespan subsists.

4.2   Rule-Support Concepts

This set of concepts has been designed for supporting reasoning activities and for tracking potential rule violations caused by users.

The concept "Rule" represents the rules defined by domain experts through the platform interface. In the current version of the ontology, instances of this concept are represented by a string composed by a set of atomic logical clauses.

The second concept, "Violation", is exploited for instantiating user's violations with respect to each rule. Such a class is defined with restrictions as following:

```
<owl:Class rdf:about="perkapp:Violation">
  <rdfs:subClassOf>
    <owl:Class>
      <owl:intersectionOf rdf:parseType="Collection">
        <owl:Restriction>
          <owl:onProperty rdf:resource="perkapp:hasViolationRule"/>
          <owl:onClass rdf:resource="perkapp:MonitoringRule"/>
          <owl:qualifiedCardinality
            rdf:datatype="&xsd;nonNegativeInteger">1
          </owl:qualifiedCardinality>
        </owl:Restriction>
        <owl:Restriction>
```

```
          <owl:onProperty rdf:resource="perkapp:hasViolationUser"/>
          <owl:onClass rdf:resource="perkapp:User"/>
          <owl:qualifiedCardinality
            rdf:datatype="&xsd;nonNegativeInteger">1
          </owl:qualifiedCardinality>
        </owl:Restriction>
        <owl:Restriction>
          <owl:onProperty rdf:resource="perkapp:hasViolationLevel"/>
          <owl:qualifiedCardinality
            rdf:datatype="&xsd;nonNegativeInteger">1
          </owl:qualifiedCardinality>
          <owl:onDataRange rdf:resource="&xsd;integer"/>
        </owl:Restriction>
      </owl:intersectionOf>
    </owl:Class>
  </rdfs:subClassOf>
</owl:Class>
```

Beside their use in reasoning activities, violation instances may be used by external services for data aggregation purposes.

4.3   User Information

Users are represented in the ontology for linking purposes only. Indeed, users' personal data are managed by external organizations, e.g., hospital institution in the case of users under medical control. Within the ontology, we modeled only the "User" concept that is used as a bridge between ontology data and actual user's information stored in the external resources. Such a concept has only a datatype property containing the unique identifier of the user within the external resources. Within the ontology, user instances are adopted for associating meals and, eventually, violations to the respective instances.

5   The PerKApp Ontology in Action

The PerKApp project does not aim to just provide yet another ontology and application wellness-related like many others already present on the market. The PerKApp project was born with the main aim of supporting different monitoring activities from sponsoring healthier lifestyles to avoiding chronic diseases and preventing onset of pathologies. Below, we will describe the role of the PerKApp ontology for supporting different tasks concerning the development of the full-fledged PerKApp platform.

Definition of Rules Schemata. As introduced above, one of the objectives on the PerKApp project is to monitor eating habits of users in order to prevent different type of pathologies and chronic diseases. Such a monitoring activity is performed by implementing within the platform a machinery for defining dietary rules. Rules definition is performed by domain experts supported by easy interfaces allowing the exploitation of ontological concepts for defining both simple and complex rules.

Designed rules combined with information provided by users can be used for performing reasoning on the populated knowledge base. Results of the reasoning activity can then be exploited for carrying out further actions with the aim of redirecting users' behaviors to better lifestyles.

Ontology Enrichment for Persuasion. In Natural Language Generation (NLG) usually two levels can be identified: strategic and tactical [8]. It should be noted that most systems and approaches in NLG are based on descriptive tasks, focusing on texts which realize a single, often informative, communicative goal, as opposed to persuasive NLG where the communicative goal is usually surmounted by reasoning about the persuadee's behavior modification. Persuasive features can have an impact on both strategic and tactical levels since the effectiveness of a message can be enhanced by appropriate content selection, text planning and linguistic choices. Within PerKApp the ontology design has been driven by the needs of "persuasive" NLP reasoning.

For example at the strategic level, diverting a user from eating unhealthy foods often requires suggesting viable alternatives. Alternatives can be either found reasoning on nutrients or by encoding more abstract concepts as taste. So, for example, given a food F_1, we can found a substitute food F_2 that has similar nutrients, minus the unhealthy ones. Another direction might be the adoption of a *"similar_to"* relation given by taste similarity. This way, it would be able to find a food F_2 that is connected to F_1 on the basis of criteria different from a mere data similarity.

APIs Service. Finally, the PerKApp ontology is exposed as a web service available on the PerKApp project website[2]. Currently, it supports three methods described below. Here, we reported only the description of each method, while, details concerning output data structures are reported on the project website.

"**/GetFoodList**": This method returns the list of foods contained within the ontology with their labels.

"**/SingleFoodData**": This method returns all information associated with a specific food. The method supports two parameters:

- *food* (mandatory): the "id" of the food the user wants to retrieve;
- *quantity* (optional): the quantity, expressed in grams, that has to be used by the service for computing the amount of nutrients contained in the requested food. Default value is 100.

"**/CheckMeal**": The last method currently implemented allows to compute the amount of calories and nutrients consumed in a given meal. This method expects two mandatory parameters:

- *mealFoods*: a string listing food ids separated by ";".
- *mealFoodsQuantity*: a string of numbers indicating, for each of the food listed within the "mealFoods" parameter, the amount of grams consumed. Also for this method, the separator between numbers is ";". Number placed at position n in this parameter corresponds to the quantity of the n-th food listed within the "mealFoods" parameter.

---

[2] The base url of the web service is "http://shellvm1.fbk.eu:8080/virtualcoach_webservice".

6   Conclusions and Future Work

In this paper, we presented the first version of the PerKApp ontology and its role in the PerKApp platform. Concepts modeled into the ontology will not have only descriptive purposes, but they will also act as support for further activities concerning the monitoring and the persuasion of patients and citizens for pursuing healthier lifestyles.

Future work on the ontology will concern the integration of further domains (e.g. the physical activity one) and the linking with existing environmental ontologies (e.g. locations, weather, etc.) that, in smart contexts, would support the development of more intelligent platforms.

References

1. Bailoni, T., Dragoni, M., Eccher, C., Guerini, M., Maimone, R.: PerKApp: A context aware motivational system for healthier lifestyles. In: ISC2, pp. 1–4. IEEE (2016)
2. Grafenauer, S., Tapsell, L., Beck, E.: Beyond nutrients: classification of foods to identify dietary patterns for weight management. In: 16th International Congress of Dietetics
3. Cantais, J., Dominguez, D., Gigante, V., Laera, L., Tamma, V.: An example of food ontology for diabetes control. In: Proceedings of the International Semantic Web Conference 2005 Workshop on Ontology Patterns for the Semantic Web (2005)
4. Snae, C., Bruckner, M.: Foods: a food-oriented ontology-driven system, pp. 168–176 (2008)
5. Helmy, T., Al-Nazer, A., Al-Bukhitan, S., Iqbal, A.: Health, food and user's profile ontologies for personalized information retrieval. In: Shakshuki, E.M. (ed.) Proceedings of the 6th International Conference on Ambient Systems, Networks and Technologies (ANT 2015), the 5th International Conference on Sustainable Energy Information Technology (SEIT-2015), London, 2–5 June 2015. Procedia Computer Science, vol. 52, pp. 1071–1076. Elsevier (2015)
6. Çelik Ertuğrul, D.: FoodWiki: a mobile app examines side effects of food additives via semantic web. J. Med. Syst. **40**(2), 1–15 (2016)
7. Gangemi, A., Guarino, N., Masolo, C., Oltramari, A., Schneider, L.: Sweetening ontologies with DOLCE. In: Gómez-Pérez, A., Benjamins, V.R. (eds.) EKAW 2002. LNCS (LNAI), vol. 2473, pp. 166–181. Springer, Heidelberg (2002). doi:10.1007/3-540-45810-7_18
8. Reiter, E., Dale, R.: Building Natural Language Generation Systems. Cambridge University Press, Cambridge (2000)

An Experimental Evaluation of Automatically Generated Multiple Choice Questions from Ontologies

Ghader Kurdi$^{(\boxtimes)}$, Bijan Parsia, and Uli Sattler

School of Computer Science, The University of Manchester,
Kilburn Building, Oxford Road, Manchester M13 9PL, UK
{ghader.kurdi,bijan.parsia,Ulrike.Sattler}@manchester.ac.uk

Abstract. In order to provide support for the construction of MCQs, there have been recent efforts to generate MCQs with controlled difficulty from OWL ontologies. Preliminary evaluation suggests that automatically generated questions are not field ready yet and highlight the need for further evaluations. In this study, we have presented an extensive evaluation of automatically generated MCQs. We found that even questions that adhere to guidelines are subject to the clustering of distractors. Hence, the clustering of distractors must be realised as this could affect the prediction of difficulty.

1 Introduction

Multiple Choice Questions (MCQs) are a widely adopted form of question in both paper- and electronic-based tests. A great proportion of large scale tests consist of MCQs. They have gained further importance with the advent of e-learning and Massive Open Online Courses (MOOCS) (e.g. Coursera, Future Learn, and Udacity), in which providing assessment and feedback on a large scale is challenging. However, MCQs are labour intensive, time consuming and difficult to construct. Well-constructed MCQs require a considerable time for design, writing, and revision. In order to provide support for the construction of MCQs, there have been recent efforts to generate MCQs with controlled difficulty from OWL ontologies based on the similarity theory of difficulty [1]. The similarity theory associates difficulty with the degree of similarity between the key (correct option) and the distractors (incorrect options). Despite the advances in the method, preliminary evaluation suggests that generated questions are not field ready yet and highlight the need for more extensive evaluation of the questions.

The objective of this study is to evaluate the quality of, and to categorise various problematic phenomena of, automatically generated MCQs from ontologies based on the aforementioned theory [1]. Another objective of this study is to distinguish issues that are intrinsic to similarity theory from natural language and presentation issues. The specific questions driving this study are:

M. Dragoni et al. (Eds.): OWLED-ORE 2016, LNCS 10161, pp. 24–39, 2017.
DOI: 10.1007/978-3-319-54627-8_3

1. What are the issues presented in automatically generated question? And to what degree are they prevalent?
2. Are these issues intrinsic properties of the similarity theory as opposed to natural language and presentation issues?

The main contribution of this study is the identification of a new problematic phenomenon of clustered distractors that influences the prediction of difficulty.

2   Materials and Methods

Experimental Data. Our study used two domain ontologies for the evaluation: the Knowledge Acquisition (KA) ontology and the Java ontology. The KA and Java ontologies were handcrafted with the purpose of question generation in mind.[1] The reason behind choosing these two ontologies is the availability of corresponding courses provided by the School of Computer Science at the University of Manchester. The ontological statistics are provided in Table 1.

Experimental Set-up. The following machine has been used to carry out the experiment presented in this study: Intel core i7 2.4 GHz processor, 8 GB RAM, running Windows OS 8.1 (HP Spectre 2015 model).

Table 1. Statistics for the experimental ontologies.

Ontology	Classes	Properties	Individuals	Logical axiom
KA	151	7	0	254
Java	305	74	0	554

2.1   MCQ Generation

We used the MCQ generator developed by Alsubait et al. [1] to automatically generate MCQs using the aforementioned ontologies as inputs. The tool generates six types of questions that are explained in Appendix A. The generated questions are classified by the tool into 'easy' or 'difficult' questions. Each question consists of a stem (a text that poses the question), a key, and a non-empty set of distractors minimally containing two distractors. Different versions can be constructed from the suggested questions by selecting different subsets of the distractors. The number of generated questions is provided in Table 2. Generating questions from the Java ontology took 12 days while generating questions from the KA ontology took around 12 h. It is clear from the table that the number of difficult questions (67 difficult questions) is low compared to the number of easy questions (2090 easy questions). The reason is that few distractors with a very high similarity to the key can be found in ontologies [1].

---

[1] For a detailed description of both ontologies, the reader is referred to [1,2].

Although the size of the Java ontology is about double the size of the KA ontology, the number of easy questions generated from the Java ontology is about 11 times larger than the number of questions generated from the KA ontology. In addition, generating questions from the Java ontology took much more time than generating questions from the KA ontology. We expect that the magnitude of the difference between the number of questions and the generation cost in terms of time is related to the depth of the inferred class hierarchy. Looking at both ontologies, we noticed that the class hierarchy of the Java ontology is divided into eleven levels compared to five levels in the KA. In addition, many classes in the Java ontology have multiple direct subsumers while classes in KA have only one direct subsumer. To illustrate the effect of this, let us consider two classes: class (A) which is located at level 11 and has two direct subsumers throughout the hierarchy, and class (B) which is located at level 5 and has a single subsumer at each level. Taking the question category "What is X" as an example, the number of generated questions for class (A) is expected to be about $2^n - 2 = 2^{11} - 2 = 2046$ questions where n represents the number of levels. However, the number of questions for class (B) is only 30. Note that to generate the questions, similar distractors for each class must be found first. This non-linear growth suggests ontologies as a supplier that can satisfy the demand for a large number of questions since adding a few classes and submission relations increases the number of generated question significantly.

2.2   Sample Selection

Due to the large number of easy questions generated, we used a stratified sampling method in which questions were divided into groups according to the question category. With regards to easy questions, we randomly selected the questions from the different groups in proportion to their number, taking into account a 95% confidence level and 5% margin of error. As the number of difficult questions was small, we evaluated them all. The total number of evaluated questions is 506 questions (67 difficult questions and 439 easy questions), as shown in Table 2.

Table 2. Statistics for the number of generated questions. Note that the sizes of the samples of easy questions are represented between parentheses.

Question category	Java		KA	
	Easy	Difficult	Easy	Difficult
Generalisation: What is X	393 (66)	6	11 (8)	0
Generalisation 2: What is X2	0	0	56 (39)	8
Specification: Which is X	260 (43)	22	15 (11)	0
Specification 2: Which is X2	88 (15)	11	82 (58)	0
Definition: Which term	207 (35)	20	2 (1)	0
Recognition: Which is odd	976 (163)	0	0	0
Total	1924 (322)	59	166 (117)	8

2.3 Evaluation Criteria

We performed a preliminary evaluation of automatically generated questions and observed some problematic questions. We then referred back to the literature that discussed and suggested guidelines for developing MCQs [3,4]. Haladyna et al. [3] conducted a review of MCQ writing guidelines for assessment. In addition, Pho et al. [4] performed an analysis of multiple choice question corpus in order to define distractor characterisation. The initial criteria for our evaluation started with suggestions described in the aforementioned studies. Table 3 gives an overview of the initial set of criteria. A detailed discussion of each criterion will be provided in the associated result section for clarity. Examples of generated questions that do not adhere to guidelines can be found in Appendix B. Then, through an iterative process of evaluating the questions, we developed a new criterion for selecting distractors that was not mentioned in the literature, as will be discussed in Sect. 3.5.

Table 3. The predefined criteria for assessing automatically generated questions (adapted from [3,4]).

Quality criterion
(Q1) The question is grammatically correct
(Q2) The question contains no clues to the key
(Q3) Options are homogeneous in grammatical structure
(Q4) Options are homogeneous in content

3 MCQ Evaluation: Results and Discussion

3.1 Grammatical Correctness

The grammatical correctness of questions is an important consideration when constructing MCQs since grammatical inconsistency could give test takers without sufficient knowledge a clue to the correct answer. In order to investigate the grammatical correctness of automatically generated MCQs, we classified questions based on the level of the grammatical corrections required into:

(MIN) minor correction: involves adding appropriate articles, fixing any subject-verb disagreement and tokenising the stem and the options, including segmentation, as well as processing of camel case and underscores;

(MED) medium correction: involves inserting or deleting up to three words from the stem and the options;

(MAJ) major correction: involves rephrasing of the stem or the options.

The distribution of questions according to the level of the grammatical corrections required is shown in Table 4. Although the majority of MCQs required only minor corrections, there is a considerable number of questions requiring major corrections. Presenting questions in OWL syntax is the main reason behind the need for major grammatical corrections. However, this issue is repairable by employing one of the available ontology verbalisers. Evaluating different verbalisers in order to choose the most suitable for the purpose of question verbalisation is a part of future work. In addition, the issues of segmentation and processing of camel-case and underscore can be achieved by employing regular expressions. The total number of questions requiring major correction was higher in the KA ontology because a higher number of questions containing sub-expressions was generated from the KA ontology (Table 11).

Table 4. Results for question evaluation in regards to the required level of grammatical corrections.

Question category	Easy			Difficult		
	Minor	Medium	Major	Minor	Medium	Major
What is X	70	4	0	6	0	0
What is X2	0	0	39	0	0	8
Which is X	54	0	0	22	0	0
Which is X2	0	0	73	0	0	11
Which term	36	0	0	20	0	0
Which is odd	159	0	4	–	–	–
Total	319	4	116	48	0	19

3.2 Syntactic Clues

One of the MCQ writing guidelines in regards to writing the choices is to avoid "choices identical to or resembling words in the stem" [3]. Alsubait et al. [1] identified word clues as a problem that affects the accuracy of the difficulty prediction. We have considered different possible similarities in wording between the stem and the options:

(SK) shared word(s) or phrase between the stem and the key;
(SD) shared word(s) or phrase between the stem and one or more distractors;
(SKD) shared word(s) or phrase between the stem and the options including the key and one or more distractors.
(ANT) a word in the stem has an antonym in one or more of the distractors.

The form (SK) should be avoided because it makes the key stand out as the correct answer. On the other hand, if word(s) or a phrase in the stem are repeated in the distractor(s) only, this make the distractor(s) more attractive to low information students. This form (SD) can be desirable because it improves the functionality of the clued distractor(s) and possibly the discrimination of the item. However, the attractiveness of the clued distractors tends to decrease the functionality of the other distractors. Finally, regarding the third form (SKD), there is a preference over other options for options that share similar wording with the stem, as mentioned earlier. This leads to the nonfunctionality of some of the distractors and increases the guessability of the item. However, we did not consider questions where all distractors share word(s) with the key and the stem as containing a syntactic clue. We identified another form of syntactic clue in which a word in the stem has an antonym in one or more of the distractors. This form also needs to be avoided because the distractor(s) are clued as the wrong answer(s). A lexical database such as WordNet can be used to acquire the antonyms of concepts in the stem. The acquired terms can be associated with the stem and taken into account during the question generation.

Table 5 shows the distribution of the evaluated questions in regards to the aforementioned forms. Table 11 shows the proportion of questions that contain syntactic clues to the total number of questions in each ontology. The evaluation indicated that 25.4% and 12.5% of difficult questions generated from the Java and the KA ontologies respectively contain clues to the keys which, in turn, make the questions easy. One of the suggested solutions is to provide alternative names using OWL annotation properties which can be used by the tool if wording similarity between the stem and key is detected.

Table 5. Results for question evaluation in regards to syntactic clues.

Question category	Easy					Difficult				
	SK	SD	SKD	ANT	No clue	SK	SD	SKD	ANT	No clue
What is X	4	27	6	0	37	1	4	1	0	0
What is X2	13	2	5	0	19	1	0	0	0	7
Which is X	15	12	6	5	19	8	1	1	0	12
Which is X2	13	8	8	0	44	2	0	0	0	9
Which term	1	13	16	2	6	4	7	7	0	2
Which is odd	0	0	0	0	163	–	–	–	–	–
Total	42	62	48	7	274	16	12	9	0	30

3.3 Syntactic Consistency

One of the recommendations from the literature regarding the syntactic structure of the options is to "keep choices homogeneous in content and grammatical structure" [3]. Another related recommendation is to avoid "grammatical inconsistencies that cue the test-taker to the correct choice" [3]. In order to investigate to what extent automatically generated questions follow these rules, we automatically annotated the distractors with syntactic information about parts of speech (i.e. nouns (NN), verbs (VB), determiner (DT), etc.) using the Stanford part-of-speech tagger[2]. We then manually applied corrections where needed to the assigned part of speech for each distractor. We compared the key and each distractor in terms of their syntactic structures independently of their meaning as suggested in [4]. We consider the distractor and the key to be:

(GC) grammatically consistent: if their assigned parts of speech are identical,
(PC) partially consistent: if they share some parts of speech,
 (IC) grammatically inconsistent: if their assigned parts of speech are totally different.

Looking at different generated questions where syntactic inconsistency presents, we concluded that grammatical inconsistency can highlight the need for modification of either the questions, or the names used in the ontology, even though this is not always associated with invalid distractors.

The number of questions that contain syntactic inconsistency and a detailed analysis of the number of syntactically consistent and inconsistent distractors is presented in Table 6. Table 11 show the percentage of distractors that are syntactically inconsistent with the key in each ontology. The proportion of distractors distributed over the three categories seems to be consistent in the two ontologies.

Table 6. Results of evaluating syntactic consistency. Note that the upper part reports the number of questions while the lower part reports the number of distractors.

Question category	Easy		Difficult	
	GC and PC	IC	GC and PC	IC
What is X	24	50	6	0
What is X2	39	0	7	0
Which is X	34	20	22	0
Which is X2	73	0	11	0
Which term	18	18	17	3
Which is odd	151	12	–	–
Total	339	100	63	3
	GC	PC	IC	GC	PC	IC
What is X	556	3,984	374	0	38	0
What is X2	45	74	0	12	39	0
Which is X	259	801	221	23	88	0
Which is X2	69	280	0	11	63	0
Which term	281	765	81	39	86	4
Which is odd	138	452	61	–	–	–
Total	1,348	6,356	737	85	314	4

---

[2] Downloaded from: http://nlp.stanford.edu/software/tagger.shtml.

3.4 Semantic Homogeneity

The guidelines suggest maintaining the homogeneity of options in MCQs (Q4 in Table 3). Pho et al. [4] define semantically homogeneous distractors as the alternatives that "share a common semantic type (expected by the question)". We observed that there are some questions for which the semantic type is deducible from the stem which, in turn, enforces the use of semantically homogeneous options. Otherwise, distractors are ruled out because of type mismatch between the distractors and the key. Based on this, we consider a distractor to be either:

(HOMO) homogeneous: if its type is compatible with the expected type of the key,

(HETERO) heterogeneous: if its type is not compatible with the expected type of the key.

We conducted an analysis by checking whether the expected answer type is suggested in the question either explicitly or implicitly. Then, we checked the compatibility of distractors with the expected answer type. Table 7 shows the results of investigating the compatibility of automatically generated MCQs with the semantic homogeneity rule (Q4). Table 11 shows the distribution of questions per ontology according to semantic homogeneity.

Table 7. Results for question evaluation in regards to semantic homogeneity.

Category	Easy			Difficult		
	Homo	Hetero	Not applicable	Homo	Hetero	Not applicable
What is X	3	0	71	0	0	6
What is X2	0	0	39	0	0	8
Which is X	12	12	30	5	0	17
Which is X2	0	0	73	0	0	11
Which term	17	18	1	14	6	0
Which is odd	0	0	163	–	–	–
Total	32	30	377	19	6	42

3.5 Clustered Distractors

All aforementioned flaws are regarded as linguistic or presentation issues that can be repaired by incorporating existing natural language processing and generation techniques. However, we observed an interesting phenomenon of the existence of interrelations between distractors in automatically generated questions. We called this phenomenon "clustered distractors". The following examples illustrate this phenomenon. The first two examples represent different versions of the same question where the difference is in the distractor sets. In the first version, distractors (A) and (B) are clustered because they both represent relational operators. A test taker who knows that relational operators are binary operators will easily eliminate the distractors and arrive at the correct answer. Hence, the item functions as a true-false question. Recognising one as a binary operator and the relation between the distractors gives a clue to the answer. However, in the second version, a test taker must consider each distractor and recognise it as a binary operator in order to arrive at the correct solution.

Stem: Which of the following is [a] Unary Operator

A. Less than or equal	A. Less than or equal
B. Less than	B. Logical OR
C. Logical complement operator	C. Logical complement operator
▲ **Key**	▲ **Key**

Another form of clustered distractors is presented in the following example generated from the Java ontology. Recognising that a primitive type and a scalar represent the same concept clue the test taker to select array because (A) and (B) cannot both be correct as MCQs require only one correct option. More examples are presented in the appendix.

Stem: Which of the following is [a] Reference Type?
  A. Primitive type
  B. Scalar
  C. Array ◄ **Key**

We define clustered distractors as a subset of distractors with very high similarity among them. Our assumption is that clustered distractors make questions easier than expected. That is, even if the question is predicted to be difficult because of the high similarity between the key and the distractors, the high similarity between the distractors draws a boundary between the key and the cluster of distractors. However, the similarity theory is blind to this fact since only the similarity between the key and distractors is considered.

The results of analysis for clustered distractors are presented in Tables 8 and 11. The evaluation indicated that the phenomenon is dominant. A considerable

Table 8. Statistics for the number of questions containing clustered distractors.

Question category	Easy		Difficult	
	Clustered	Not clustered	Clustered	Not clustered
What is X	71	3	6	0
What is X2	14	25	0	8
Which is X	52	2	22	0
Which is X2	43	30	11	0
Which term	26	10	13	7
Which is odd	163	0	–	–
Total	369	70	52	15

Table 9. Statistics for the number of flawed questions and the level of repair required.

Category	Easy			Difficult		
	Flawless	1 flaw	≥2 flaws	Flawless	1 flaw	≥2 flaws
What is X	0	5	69	0	0	6
What is X2	0	14	25	0	7	1
Which is X	0	0	54	0	0	22
Which is X2	0	13	60	0	0	11
Which term	5	0	31	0	3	17
Which is odd	20	130	13	–	–	–
Total	25	162	252	0	10	57
	None	MIN	MED	MAJ	None	MIN	MED	MAJ
What is X	0	63	6	5	0	4	1	1
What is X2	0	20	6	13	0	7	0	1
Which is X	0	26	5	23	0	11	4	7
Which is X2	0	53	9	11	0	9	0	2
Which term	0	23	11	2	0	14	5	1
Which is odd	25	138	0	0	–	-	-	-
Total	25	323	37	54	0	45	10	12

Table 10. The proportion of flawed questions per ontology.

Difficulty	Category	Java		KA	
		Number	Percentage	Number	Percentage
Easy	Flawless	25	7.76%	0	0
	1 flaw	136	42.24%	26	22.22%
	≥2 flaws	161	50%	91	77.78%
Difficult	Flawless	0	0	0	0
	1 flaw	3	5.09%	7	87.50%
	≥2 flaws	56	94.92%	1	12.50%
The level of repair required					
Easy	Not required	25	7.76%	0	0
	Minor	259	80.44%	64	54.70%
	Medium	19	5.90%	18	15.39%
	Major	19	5.90%	35	29.92%
Difficult	Not required	0	0	0	0
	Minor	38	64.41%	7	87.50%
	Medium	10	16.95%	0	0
	Major	11	18.64%	1	12.50%

Table 11. The proportion of questions per ontology distributed according to: (A) grammatical corrections, (B) syntactic clues, (C) syntactic consistency, (D) semantic homogeneity, and (E) clustered distractors

Difficulty	Category	Java		KA	
		Number	Percentage	Number	Percentage
(A) Grammatical corrections					
Easy	Minor	299	92.86%	20	17.09%
	Medium	4	1.24%	0	0
	Major	19	5.90%	97	82.91%
Difficult	Minor	48	81.36%	0	0
	Medium	0	0	0	0
	Major	11	18.64%	8	100%
(B) Syntactic clues					
Easy	SK	20	6.2%	22	18.80%
	SD	50	15.5%	12	10.26%
	SKD	28	8.7%	20	17.09%
	ANT	7	2.2%	0	0
	No clue	222	68.9%	52	44.44%
Difficult	SK	15	25.4%	1	12.5%
	SD	12	20.3%	0	0
	SKD	9	15.3%	0	0
	ANT	0	0%	0	0
	No clue	23	39%	7	87.5%
(C) Syntactic consistency (no. of questions)					
Easy	GC and PC	231	71.74%	108	92.31%
	IC	91	28.26%	9	7.69%
Difficult	GC and PC	56	94.92%	7	100%
	IC	3	5.09%	0	0
(C) Syntactic consistency (no. of distractors)					
Easy	GC	1,258	15.7%	91	17.95%
	PC	6,028	75.6%	369	72.78%
	IC	690	8.6%	47	9.27%
	Total	7,976	100%	507	100%
Difficult	GC	73	20.74%	3	9.68%
	PC	275	78.13%	28	90.32%
	IC	4	1.14%	0	0
	Total	352	100%	31	100%
(D) Semantic homogeneity					
Easy	Homogeneous	23	7.14%	9	7.69%
	Heterogeneous	30	9.33%	0	0
	Not applicable	269	83.54%	108	92.31%
Difficult	Homogeneous	19	32.20%	0	0
	Heterogeneous	6	10.17%	0	0
	Not applicable	34	57.63%	8	100%
(E) Clustered distractors					
Easy	Clustered distractors	305	94.72%	64	54.70%
	Not clustered distractors	17	5.28%	53	45.30%
Difficult	Clustered distractors	52	88.14%	0	0
	Not clustered distractors	7	11.86%	8	100%

number of questions in both ontologies contain clustering of distractors, with the Java ontology having a higher percentage (94.7% of easy questions and 88.1% of difficult questions). All questions in the question category "Which is odd" contain clustered distractors, which is the nature of this category of questions. One of the patterns that we noticed with regards to clustered distractors is that they represent siblings in the ontology. This is not surprising as it is expected that, in ontologies, siblings are usually very similar to each other.

3.6 Level of Repairs

The final phase of the evaluation was to investigate the relationship between the flaws in the questions and the effort required to repair the questions. We classified questions in terms of the level of repairs required into:

- minor repair: involves minor grammatical corrections and selecting distractors if enough distractors are provided by the tool;
- medium repair: involves medium grammatical corrections and writing one distractor in order to have a question with one key and 2 distractors (3 options MCQs) if not enough distractors are provided;
- major repair: involves major grammatical correction and writing two or more distractors in order to have a question with one key and 2 distractors (3 options MCQs) if not enough distractors are provided.

The results are summarised in Tables 9 and 10. It is not surprising that few questions are flawless given the fact that no natural language generation techniques were incorporated into the tool. Filtering flawed questions will result in an insufficient number of questions (Table 9). Although the majority of questions contain more than one flaw, most of them are repairable by applying minor repairs. This is because a large number of distractors per question is suggested.

4   Conclusion

In this study, we have presented an evaluation of automatically generated MCQs. The objective was to validate the quality of the questions and thus later be able to improve the automatic question generation process. The study confirms the need to present questions more naturally. Syntactic, and syntax-based similarity as well as semantic similarity between options must be taken into consideration when automatically selecting distractors from ontologies. Available natural language processing and generation techniques, as well as some ontology modeling guidelines, suffice to overcome the linguistic issues. Alternatively, an automatic checker would be highly valuable in highlighting problematic questions and minimising review time. We also found that even questions that adhere to guidelines are subject to the clustering of distractors. This is a significant issue that is related to the core of the generation process "the similarity theory". Although this phenomenon does not weaken the validity of the similarity theory, it highlights the need for more sophisticated application of similarity. Hence, different patterns of similarity between the options must be realised as this could affect the prediction of difficulty. We are planning to validate the effect of clustered distractors on difficulty and to develop strategies to avoid or highlight such distractors when generating questions.

Acknowledgments. The authors would like to thank Tahani Alsubait for sharing the MCQ generator code.

Appendix

A   Question Categories

Table 12. Explanation of the six question categories generated by the MCQ generator (adapted from [1]).

Question category	Stem	Key	Distractors
Generalisation	What is X? where X is an atomic concept name	an atomic subsumer of X	atomic non-subsumers of X
Generalisation 2	What is X? where X is an atomic concept name	a complex subsumer (concept expression) of X	complex non-subsumers of X
Specification	Which is X? where X is an atomic concept name	an atomic subsumer of X	non-subsumees of X excluding subsumers and siblings of the stem
Specification 2	Which is X? where X is a complex concept	an atomic subsumee of X	non-subsumees of X excluding subsumers of the stem
Definition	Which term can be defined as 'annotation'	an atomic concept name annotated with the annotation	atomic concept names not annotated with the annotation
Recognition	Which is odd?	an atomic concept name not subsumed by X where X is a concept name	atomic concept names subsumed by X

B   Example Questions

Syntactic Clues

Examples of the form (SK)

Stem: State Transition Network ...:
    A. is Produced By some Concept Map Technique
    B. is Produced By some Process Map Technique
    C. is Produced By some State Transition Technique ◄ **Key**

Examples of the form (SD)

Stem: Repertory Grid Stage 2 ...
    A. involves Providing A Running Commentary
    B. involves Repertory Grid Stage 1
    C. involves Rating Concepts Against Attributes ◄ **Key**

Stem: Which of the following terms can be defined by "a Java keyword used to declare a variable that holds an 8 bit signed integer"?
    A. Char
    B. Short
    C. Int
    D. Byte ◄ **Key**

Examples of the form (ANT)

Stem: Which of the following is [a][3] Binary Operator?
    A. Unary operator
    B. Unary minus operator
    C. Equality operator ◄ **Key**
    D. Logical complement operator

Syntactic Consistency

Stem: What is [a] Book?
    A. (Is)VB (A)DT
    B. (Has)VB (Part)NN
    C. (Concept)NN ◄ **Key**

Stem: Which of the following is [a] Java Language Feature?
    A. (Recursion)NN ◄ **Key**
    B. (Implementation)NN
    C. (Requirement)NN (analysis)NN
    D. (Throw)VB

---

[3] This is a grammatical correction that is manually added to the question.

Note that in the previous example, although (D) is inconsistent with the key, it is indeed a Java language feature.

Stem: Which of the following terms can be defined by "A binary remainder operator that produces a pure value that is the remainder from an implied division of its operands"?

    A. (Divide)VB

    B. (Multiply)VB

    C. (Modulus)NN ◄ **Key**

In this example, both distractors (A) and (B) are inconsistent with the key but they are both plausible. By investigating the ontology, we found that this issue resulted from the inconsistent naming of concepts.

Semantic homogeneity

Stem: Which of the following terms can be defined by "A layout manager that allows subcomponents to be added in up to five places specified by constants NORTH, SOUTH, EAST, WEST and CENTER"?

    A. Simple Object (heterogeneous)

    B. Event (heterogeneous)

    C. Border Layout (homogeneous) ◄ **Key**

    D. Grid Layout (homogeneous)

It is clearly deduced from the previous question that the expected answer is a layout manager. As can be seen, distractors (A) and (B) are heterogeneous in relation to the key type while option (D) is homogeneous.

Clustered Distractors

Stem: Which of the following terms can be defined by "A stage in the software development process where customer needs are translated into how it could be implemented"?

    A. Testing

    B. Unit Testing

    C. Implementation

    D. Design ◄ **Key**

Distractors (A) and (B) are clustered since knowing that the answer is not testing will allow the elimination of all types of testing.

Stem: Protocol Analysis Technique ...

    A. involves Repertory Grid Stage 1

    B. involves Repertory Grid Stage 2

    C. involves Repertory Grid Stage 4

    D. involves Identifying Knowledge Objects ◄ **Key**

Stem: Which of the following is produces some Protocol?

    A. Attribute Laddering

    B. Process Laddering

    C. Laddering

    D. Semi-structured Interview ◄ **Key**

References

1. Alsubait, T., Parsia, B., Sattler, U.: Generating multiple choice questions from ontologies: lessons learnt. In: OWLED, Chicago, pp. 73–84 (2014)
2. Alsubait, T., Parsia, B., Sattler, U.: Generating multiple choice questions from ontologies: how far can we go? In: Lambrix, P., Hyvönen, E., Blomqvist, E., Presutti, V., Qi, G., Sattler, U., Ding, Y., Ghidini, C. (eds.) EKAW 2014. LNCS (LNAI), vol. 8982, pp. 66–79. Springer, Cham (2015). doi:10.1007/978-3-319-17966-7_7
3. Haladyna, T.M., Downing, S.M., Rodriguez, M.C.: A review of multiple-choice item-writing guidelines for classroom assessment. Appl. Measur. Educ. **15**(3), 309–333 (2002)
4. Pho, V.-M., Andre, T., Ligozat, A.-L., Grau, B., Illouz, G., Francois, T., et al.: Multiple choice question corpus analysis for distractor characterization. In: LREC, pp. 4284–4291, Reykjavik (2014)

Use Cases and Suitability Metrics
for Unit Ontologies

Markus D. Steinberg, Sirko Schindler, and Jan Martin Keil[(✉)]

Heinz Nixdorf Chair for Distributed Information Systems,
Institute for Computer Science, Friedrich Schiller University Jena, Jena, Germany
{markus.daniel.steinberg,sirko.schindler,jan-martin.keil}@uni-jena.de

Abstract. Units of measurement are an essential part of dataset descriptions as they are required for a valid interpretation of the data. One obvious choice for representing units are ontologies, but as every application supports different use cases a multitude of ontologies has been created. Each of these is suited best for just a subset of the possible use cases. The problem of choosing an ontology for a new project hence consists of two major aspects: What use cases need to be covered and which ontology caters best to them?

We describe possible use cases and analyze their requirements. The results are then used to assess the modeling of the domain in different ontologies with respect to their suitability for those use cases. This analysis shows the differences in the support for different use cases. It can help developers to choose the best ontology for their specific needs and also highlights areas for further ontology improvement.

Keywords: Measurement unit · Ontology · Ontology comparison · Ontology evaluation · Use cases

1 Introduction

Units of measurement like meter, kilogram or yard are essential for a precise description of data. They alone allow an unambiguous interpretation of values in datasets. Ontologies like the ones proposed in [1–5] provide one good option for modeling this aspect. However, different data-centric applications cater to different audiences and provide different functionalities. As a consequence, ontologies created by different projects differ in their level of support for individual use cases. This situation challenges new projects to select a suitable ontology that fits their specific needs.

In this paper, we aim to provide support for such a decision, by analyzing a set of use cases for unit of measurement ontologies. After providing the necessary background in Sect. 2, we present possible use cases in Sect. 3. In order to cover all relevant aspects, we have complemented use cases described in the literature with a number of new ones. Following this requirements analysis, a set of seven

The original version of this chapter was revised: this chapter was previously published non-open access. The correction to this chapter is available at
https://doi.org/10.1007/978-3-319-54627-8_12

M. Dragoni et al. (Eds.): OWLED-ORE 2016, LNCS 10161, pp. 40–54, 2017.
DOI: 10.1007/978-3-319-54627-8_4

ontologies will be studied to check their support for each requirement. In Sect. 4 suitable metrics will be defined to rate each ontology use case pairing. Finally, in Sect. 5 the ontologies are evaluated with respect to their suitability for each use case. This results in a ranking of ontologies for each use case, which can be used to identify the best existing ontology for new projects' use cases.[1]

2   Related Work

Several use cases (UCs) for unit ontologies are described in [2,3,5,7–9]. They will be reviewed in detail in Sect. 3. In [8] the coverage of features in multiple unit ontologies was analyzed. This analysis determined a lack of a unit ontology containing all important concepts of this domain. In [7] five feature support levels were defined to rank unit ontologies, which provides a fast overview of scope and level of development of ontologies. The order of requirements for each ranking level, however, seems biased by the author's background. An example is conversions, which are necessary to the second level. Even an ontology modeling all other features mentioned can not go beyond level two as long as it is missing conversions. Finally, the ranking was applied to multiple ontologies. Nevertheless, a metric based suitability evaluation of unit ontologies per UC is still missing.

The application of Competency Questions (CQs) [10] is a popular method in the field of ontology engineering to describe the required concepts for a UC of an ontology, that can also be used for ontology evaluation [11]. However, this approach is limited to the mere assessment of a single ontology, instead of comparing multiple ones. Furthermore, if the list of requirements can be gathered otherwise, it is not mandatory to formulate CQs. Therefore, a metric that directly uses a list of requirements is favorable.

OntoQA [12] is a popular set of metrics in the field of ontologies. These metrics provide different relationship based rankings of a schema and its classes and instances. In addition, it is possible to provide a keyword list, to focus the ranking on relevant terms. But a high ranking does not assure that an ontology can fulfill a given UC, even if an adequate keyword list was provided.

Another extensive set of metrics is provided by OntoMetric [13]. It consists of a taxonomy of 160 metrics in the five main branches *content, language, methodology, tools* and *costs* and a method to calculate the total ranking of the ontologies. This includes, for instance, the metrics *essential concepts* and *essential relations*. The metrics can be weighted by the user, but it is not possible to rank the importance of the required concepts and relations.

3   Use Cases

To evaluate the suitability of ontologies for a certain use case, the corresponding requirements have to be known. Therefore we will provide a description and a requirements analysis for each use case. Requirements will be distinguished in necessary and optional requirements. Necessary requirements are features that

---

[1] The terminology throughout this paper follows the definitions given in [6].

make the ontology eligible for a use case - if one of them is not modeled, the ontology is not able to provide even basic support for the use case. Optional requirements are those that are not necessary but simplify the implementation of a use case or increase its usefulness. Besides covering all use cases mentioned in the literature, we also provide some new use cases (marked by *) that have, to the best of our knowledge, not yet been presented. To provide a better overview, we group use cases that are concerned with similar domains.

Figure 1 outlines the use case grouping, while Table 1 summarizes the relationship between use cases and requirements.

Group 1 (Data Annotation). The first group consists of use cases that are related to data annotation. Data annotation here is the assignment of a unit of measurement or kind of quantity to a dataset or parts thereof. Consistent and consequent data annotation can prevent misunderstandings and ambiguities when exchanging, merging or comparing datasets.

UC 1 (Manual Annotation). [2,7]

An ontology can assist manual data annotation by providing lists containing kinds of quantities or units of measurement for the user to choose from.

Example: Before publishing a dataset, researchers have to create meta data, which includes annotation with units of measurement.

Necessary: An ontology has to model *kinds of quantities* or *units of measurement.*

Optional: The *connection between kinds of quantities and units of measurement* can be modeled so after choosing from one list, the other one is limited

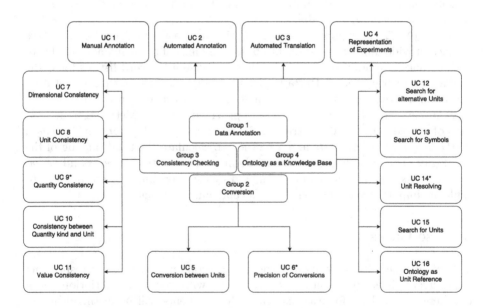

Fig. 1. Schematic overview over use cases and groups.

Table 1. Requirements per use case. (●...necessary requirement; ◐...optional requirement; ○...not required; for entries with the same index at least one has to be present)

	unit	kind of quantity (qk)	field of application (app)	dimension (dim)	system of units (system)	phenomenon (phen)	measurement (meas)	conversion (conv)	prefix (prefix)	unit ↔ system	unit ↔ qk	unit ↔ dim	unit ↔ vector	unit ↔ prefix	unit ↔ app	qk ↔ app	meas ↔ phen	meas ↔ qk	meas ↔ unit	meas ↔ value	symbols for units	symbols for qks	typ. values per qks and apps	typ. values per units and apps	allowed values per units	allowed values per units and apps	precision of conversion	number of conversion	unit of diff. lang. tags	quantity composition	unit composition	everyday lang. designators	resolvable URLs
Group 1 UC 1	●1	●1	○	◐	◐	○	○	○	◐	◐	○	○	○	◐	◐	○	○	○	○	○	◐	◐	◐	◐	◐	○	○	○	○	○	◐	○	○
UC 2	●2	●2	○	◐	◐	○	○	○	◐	◐	○	○	○	◐	◐	○	○	○	○	○	◐	◐	◐	◐	◐	○	○	○	◐	○	◐	◐	○
UC 3	●3	◐3	○	◐	◐	○	○	○	◐	◐	○	○	○	◐	◐	○	○	○	○	○	◐	◐	◐	◐	●	○	○	○	○	○	◐	◐	○
UC 4	●	○	○	○	◐	○	○	●	◐	●	○	○	○	◐	◐	○	●	●	●	●	◐	◐	◐	◐	◐	○	○	○	○	○	◐	◐	○
Group 2 UC 5	●	○	○	○	○	●	○	○	○	○	○	○	○	○	○	○	○	○	○	○	○	○	○	○	○	●	○	○	○	○	○	○	○
UC 6	●	○	○	○	○	●	○	○	○	○	○	○	○	○	○	○	○	○	○	○	○	○	○	○	●	○	○	○	○	○	○	○	○
Group 3 UC 7	●	○	○	○	○	○	○	○	○	○	○	○	○	○	○	○	○	○	○	○	○	○	○	○	○	○	○	●	○	○	○	○	○
UC 8	●	○	○	○	○	○	○	○	○	◐	○	○	○	○	○	○	○	○	○	○	○	○	○	○	○	○	○	○	○	○	○	○	○
UC 9	○	○	●	○	○	○	○	○	○	◐	○	○	○	○	○	○	○	○	○	○	○	○	○	○	●	○	●	○	○	○	○	○	○
UC 10	●	●	○	○	○	○	○	●	○	◐	○	○	○	○	○	○	○	○	○	○	◐	○	○	○	○	○	●	○	○	○	○	○	○
UC 11	●4	●4	○	○	○	○	○	○	○	○	○	○	○	○	○	○	○	○	○	○	◐	○	5	◐5	●	○	○	○	○	○	◐	○	○
Group 4 UC 12	●6	◐6	◐6	○	○	○	○	○	○	●7	7	●7	○	○	○	6	○	○	○	○	◐	◐	◐	◐	○	○	○	○	○	○	◐	◐	○
UC 13	●8	◐8	○	○	○	○	○	○	○	7	7	○	○	○	○	○	○	○	○	○	◐	◐	9	9	○	○	○	○	○	○	◐	◐	○
UC 14	●	○	○	○	○	○	○	○	○	○	○	○	○	○	○	○	○	○	○	○	○	◐	9	9	○	○	○	○	○	○	◐	◐	○
UC 15	●	◐	○	○	○	○	○	○	○	◐	◐	○	○	○	○	○	○	○	○	○	◐	◐	○	○	◐	○	○	○	○	○	◐	○	○
UC 16	●	◐	○	○	○	○	○	○	○	◐	◐	○	○	○	○	○	○	○	○	◐	○	○	○	○	◐	○	●	○	○	○	◐	○	●

to matching entries. In the same way, *fields of application and their connections to units of measurement or kinds of quantities* as well as *systems of units and their connections to units* can be used. Additionally, if there are values given in the dataset, those can be used alike if there is a model of *typical or allowed values for kinds of quantities or units of measurement*. The content of the lists can also be translated into the preferred language of the user if there are *labels in multiple languages* present in the ontology. To improve the visual representation of annotated data, *symbols for units and kinds of quantities* can be included.

UC 2 (Automated Annotation). [2]

When the amount of datasets grows, manual annotation is not feasible anymore and has to be replaced by an automatic approach. An ontology can enable a system to automatically derive kinds of quantities or units of measurement from a textual description.

Example: For populating a new, semantically enhanced data management platform with a large amount of datasets, they have to be annotated.

Necessary: An ontology has to model *kinds of quantities or units of measurement* to enable this.

Optional: To improve the efficiency of such a system the ontology can include the *connection between units of measurement and kinds of quantities*. It can also model *fields of application* and *systems of units* as well as the respective *connections to units of measurement and kinds of quantities*. Additionally, *typical and allowed values per units of measurement or per kinds of quantity* can be used to limit the possible options.

The textual description can contain symbols and be written in the user's preferred language, so *models for symbols* and *labels in multiple languages* can be exploited, too. In [2] the authors also mention modeling *everyday language designators* to handle common mistakes like writing "weight" instead of "mass".

UC 3 (Automated Translation). [3,8]

Designators, e.g., for kinds of quantities and units of measurement can automatically be translated for annotated data to cater to users of different language backgrounds. This will also reduce the number of errors as a result of missing (English) language skills.

Example: When datasets are exchanged between researchers each individual can work on them using their own language.

Necessary: An ontology needs to provide *models for units of measurement or kinds of quantities* and *labels in at least two languages*.

UC 4 (Representation of Experiments). [2]

An ontology can be used to represent observations and experiments. [2] defines an observation as a link between a phenomenon, a kind of quantity, a numerical value and a unit of measurement. Hence, this can be interpreted as the annotation of an observation with the aforementioned concepts of the ontology.

Example: A user wants to represent his measurement of the height of a certain specimen within an ontology.

Necessary: An ontology needs to provide models for *units of measurement, kinds of quantities, measurements* and the *connections between those concepts*. Additionally, there has to be the possibility to state the *measured phenomenon* and the *measured value*.

Optional: The suitability can further be improved if the ontology itself models *phenomenon* so no further ontology has to be included.

Group 2 (Conversion). The second group consists of use cases that are related to conversions between units. Unit conversion is changing the unit used to represent a measurement.

UC 5 (Conversion between Units). [2,7]

For the unit conversion a proper formula has to be provided.

Example: Differences in measured units can easily be overcome as, e.g., measurements taken using imperial units can be converted into the metric system.

Necessary: An ontology has to model *units* and a *conversion* between them. A conversion here consists of a conversion factor and an offset.

UC 6 (*Precision of Conversions).

Many applications depend on exact data. Due to the limited precision of floating point arithmetics in computer systems, conversions influence the accuracy of the converted data. As a consequence, an ontology has to augment each conversion it provides with an estimation of the respective accuracy for the values.

Example: Many conversions introduce an error of some degree. For the final result of possibly multiple conversions one has to be able to estimate whether the achieved accuracy of the result still matches the given requirements.

Necessary: An ontology needs to model *units of measurement, conversion* and *information about the precision* for the latter.

Group 3 (Consistency Checking). The third group includes all use cases that check formulas or annotated terms for consistency. In [3] consistency checking is mentioned but is not described in detail. Hence, is not listed as a reference in the individual use cases.

UC 7 (Dimensional Consistency). [2,7]

Equations and terms can be checked for dimensional consistency by comparing the dimensions or dimension vectors of all its components. Individual terms can also be checked for conformance with a given dimension vector. In [7] the necessity to check code for dimensional consistency is mentioned, too.

Example: Considering a formula like "$x\ m + y\ ft = z\ pc$" a system should state that the formula is dimensional consistent.

Necessary: An ontology has to model *dimension vectors, units of measurement* and the *connection* between them to be suitable for this use case.

Optional: The suitability can be improved by modeling *dimensions* and their *connection to units of measurement* so equations do not have to be compared by their dimension vectors but their dimensions.

UC 8 (Unit Consistency). [2]

In extension of UC 7, not only the dimensions of the involved components are compared, but also the actually used units. This highlights cases, where, e.g., values given in *meter* and *foot* are added without the necessary conversions.

Example: Using the same formula as UC 7, "x m $+$ y ft $= z$ pc", a system should this time determine that the formula is not unit consistent.

Necessary: An ontology has to model *units of measurement* and *unit compositions*.

UC 9 (*Quantity Consistency).

Similar to UC 8 the consistency with regard to kinds of quantities can also be tested. An equation or term is considered quantity consistent if all its components use kinds of quantities in a compatible manner.

Example: Adding two *lengths* is considered compatible, whereas adding a *width* and a *height* is not, although they might share the same unit of measurement.

Necessary: An ontology has to model *kinds of quantities* and the *quantity composition*.

UC 10 (Consistency between Kind of Quantity and Unit of Measurement). [2]

Each kind of quantity is accompanied by a set of units of measurement that can be used to express observations of it. A system can now check for the cases, where a unit of measurement is used in conjunction with a kind of quantity without being assigned to it.

Example: A measurement of *two meters* is considered compatible to *height*, whereas a measurement of *two seconds* is not.

Necessary: To check consistency between a given *unit of measurement* and a *kind of quantity* an ontology has to provide both concepts and a *connection* between them.

UC 11 (Value Consistency). [8]

Some units of measurement and kinds of quantities have a restricted range of allowed values. A system can assure the data quality by checking entered data.

Example: A value of *minus five* for *degree Celsius* is considered compatible, whereas for *Kelvin* it is not.

Necessary: To check if values that are annotated with such a kind of quantity or unit of measurement lie within those ranges, an ontology has to model *units of measurement* or *kinds of quantities* and *the respective allowed values*.

Optional: To further improve on this, an ontology can not only model allowed values but also *typical values for units of measurement or kinds of quantities*. Since typical values vary heavily depending on the *field of application*, they should be stated *per field of application*. A model for *conversions* between units can help further, because typical and allowed values for units of measurement, that have not been specified, can then be calculated from the values of other units of measurement.

Group 4 (Ontology as a Knowledge Base). The ontology can be used as a knowledge base to search for important information. Depending on the kind of information, multiple use cases can be distinguished.

UC 12 (Search for alternative Units of Measurement). [8]

An ontology can be used to search for possible alternatives given a unit of measurement. To determine the set of possible alternatives kinds of quantities, dimensions or dimension vectors can be used.

Example: When encountering an unfamiliar unit like *Gunter's chain* this allows for easy access to possible alternatives like *meter*.

Necessary: An ontology has to model *units of measurement* and *kinds of quantities, dimensions or dimension vectors* as well as *their connections to units of measurement*.

Optional: Similar to the manual annotation, the suitability for this use case can be improved by modeling *fields of application* and *systems of units* and their *connections to units of measurement* so the number of possible alternatives can be reduced.

UC 13 (Search for Symbols). [8,9]

Symbols for units and kinds of quantities can, e.g., be used for informal data annotation or for a shortened representation in a user interface. The search for symbols and abbreviations for units of measurement or kinds of quantities therefore is an everyday use case.

Example: When creating natural language texts from more formal data sources measurements usually will use abbreviations of used units instead of their full name.

Necessary: An ontology has to model *kinds of quantities* or *units of measurement* and the respective *symbols*.

UC 14 (*Unit Resolving).

In unit resolving, one is given a formula and the unit for each contained value. The task is now to determine the resulting unit of this formula. This assumes, that the formula is consistent with regard to UCs 7 to 9.

Example: Given a formula like "$x \ kg \times y \ \frac{m}{s^2} = z$?" a system has to deduce that the missing unit could be *Newton*.

Necessary: This use case relies on *units of measurement* and *unit composition* because it has to compute possible compositions for the units of measurement used in the formula.

Optional: It can further be improved by using *conversions* so that mismatching units can automatically be converted.

UC 15 (Search for Units of Measurement). [8]

The search for units of measurement is not restricted to alternatives, but can use a variety of different inputs. The input can, for example, consist of kinds of quantities, symbols, dimensions, dimension vectors, prefixes, systems of units or any combination of those.

Example: A user is looking for a metric unit of measurement for the kind of quantity *length* that uses the prefix *kilo.*

Necessary: Any ontology that models *units of measurement* is eligible to support the search for units because a plain list is sufficient to choose a unit of measurement.

Optional: Each concept modeled in addition can improve the suitability by enabling more input combinations and therefore narrowing down the results. These concepts are *kinds of quantities, symbols for units of measurement, fields of application, dimensions, dimension vectors, prefixes, systems of units* and the *connections between each of those concepts and units of measurement.* Labels *in multiple languages* and *everyday language designators* can also be helpful in order to enable users to state input in their preferred language.

UC 16 (Ontology as Unit Reference). [3,5]
A unit ontology can be used as a reference by other ontologies by providing unique identifiers for units of measurement.

Example: An ontology about animals can reuse the definition of *meter* or *kilogram* in the description of specimen, without having to redefine them.

Necessary: An ontology only needs to model *units of measurement.*

Optional: To improve the suitability for this use case, more concepts can be modeled to provide even more unique identifiers. These concepts are *systems of units, kinds of quantities, fields of application* and *dimensions.* To enable the user to easily access further information, there should be *labels in multiple languages* and *resolvable URIs* for the ontology.

4   Methods

We will use a metric to evaluate the suitability of an ontology for a UC. This metric depends on the list of necessary and optional requirements of each UC outlined in Sect. 3. To simplify the metric we first define a set of sub-metrics. For each required concept, relation or other feature, except the language support, we define a boolean metric m in Eq. 1. Those sub-metrics remain boolean since we are only concerned with the mere existence of a feature and not the extent of its usability.

$$m = \begin{cases} 1 & : \text{concept, relation (direct or indirect) or feature contained} \\ 0 & : \text{otherwise} \end{cases} \quad (1)$$

RDF provides a dedicated mechanism for the usage of different languages by allowing developers to attach language tags to labels [14]. Hence, the ontologies do not have to model this on their own. To assess the support, we check the usage of the RDF concept. The value an ontology reaches should be the higher the more languages are supported by it. Therefore we need a metric to rate the number of different languages l in an ontology.

$$m_{lang} = 1 - \frac{1}{l+1} \quad (2)$$

Finally, we define for each UC the encompassing suitability metric m_{suit} as the aggregation of its sub-metrics:

$$M_{nec} = \{m \mid m \text{ is metric of a necessary requirement}\} \tag{3}$$

$$M_{all} = \{m \mid m \text{ is metric of a necessary or optional requirement}\} \tag{4}$$

$$m_{suit} = \left(\min_{m \in M_{nec}} \lceil m \rceil \right) \times \left(\sum_{m \in M_{all}} \frac{m}{|M_{all}|} \right) \tag{5}$$

The first part in Eq. 5 ensures that an ontology is rated with zero if at least one necessary feature is missing. The ceiling function is necessary to accommodate for the language sub-metric. The second part is the average over all sub-metrics and provides a gradation between ontologies, that implement a different number of optional requirements. All sub-metrics are equally weighted for now, but this can easily be extended to use a vector of weights.

5   Results

To evaluate the current state of ontology development in the field of units of measurement we applied the requirements of the use cases identified in Sect. 3 and the metrics defined in Sect. 4. We analyzed the following seven prominent representatives of unit ontologies.

- Measurement Units Ontology (MUO)[2]; result of a project to exploit semantics in mobile environments; the instances were automatically generated from UCUM [15],
- Extensible Observation Ontology (OBOE)[3]; an ontology suite to represent scientific observations,
- Ontology of units of Measure and related concepts (OM)[4]; an ontology to model concepts and relations important to scientific research, developed in the context of food research [2],
- Library for Quantity Kinds and Units (QU)[5]; a showcase ontology based on the OMG SysML 1.2 QUDV specifications and the UN/CEFACT Recommendation 20 code list [16],
- Quantities, Units, Dimensions and Data Types Ontologies (QUDT)[6]; developed in the context of NASA projects,

---

[2] `muo-vocab.owl` and `ucum-instances.owl` dated 2008 from
  http://idi.fundacionctic.org/muo/.
[3] Version 1.0 from https://semtools.ecoinformatics.org/oboe.
[4] Version 1.8.2 dated 2016-03-22 from
  http://www.wurvoc.org/vocabularies/om-1.8/.
[5] `qu.owl` and `qu-rec20.owl` dated 2011-06-28 from
  https://www.w3.org/2005/Incubator/ssn/ssnx/qu/.
[6] Version 1.1 from http://www.qudt.org/.

– Semantic Web for Earth and Environmental Terminology (SWEET)[7]; also developed in the context of NASA projects and
– Units of Measurement Ontology (UO)[8] + Phenotypic Quality Ontology (PATO)[9]; both modules of the OBO family to model units and phenotypic qualities.

In a first step, each ontology was examined with respect to the requirements. In the process, the results of [17,18] were used where possible. In that project we analyzed the ontologies' instances with respect to their distribution and possible errors. Bear in mind, though, that with this work we are just analyzing ontologies with regard to their basic support for use cases and not the extent of such support. As a consequence, a feature is regarded as supported if there is any modeling of such a feature. The number of actual instances of such a feature does not matter as long as there is a matching concept. Note, furthermore, that the modeling of concepts related to UC 4 like *phenomenon* or *measurement* is not part of [18] and therefore had to be checked manually.

To judge the number of languages used by an ontology we counted the number of different language tags appearing within. This, however, is not accurate as ontologies do not seem to use language tags consequently: Even if a language tag is used in the label for one instance, one should not assume the same for all instances. Sometimes the language tag is even missing entirely. That is if there is a label at all, which can not be taken for granted. To improve this sub-metric a further analysis on an instance level has to be conducted. In this work, however, the main focus was the modeling used by the ontologies and hence the number of different language tags seems a suitable approximation. The existence of features in the ontologies as per our analysis is given in Table 2.

Using the requirements of Sect. 3, the metrics presented in Sect. 4 and the results from Table 2 a suitability score has been computed for each pair of ontology and use case. Table 3 shows an overview of the computed values. Note that the sub-metric describing the presence of language tags can never reach a value of one (cp. Eq. (2)). As a consequence all metrics using that sub-metric should only be used to compare ontologies and not to rate a single ontology.

The support for different use cases varies quite a lot. One prime example is data annotation (Group 1): While both manual (UC 1) and automatic (UC 2) annotation are basic features supported by all ontologies, the translation of designators (UC 3) on the other hand oftentimes fails as just OM contains multiple languages for its labels. The representation of experiments (UC 4) fails in most ontologies as well due to missing concepts in that area.

Conversion (Group 2) in its basic form (UC 5) is supported by almost all ontologies, but no ontology includes any estimation of the accuracy of the provided values (UC 6).

---

[7] Version 2.3 from http://sweet.jpl.nasa.gov/.
[8] Version 2016-05-13 from http://purl.obolibrary.org/obo/uo.owl.
[9] Version 2016-05-22 from http://purl.obolibrary.org/obo/pato.owl.

Table 2. The presence of features within the examined ontologies. (●...feature modeled; ○ ...feature not modeled)

	MUO	OBOE	OM	QU	QUDT	SWEET	UO
unit of measurement (unit)	●	●	●	●	●	●	●
kind of quantity (qk)	●	●	●	●	●	●	●
field of application (app)	○	○	●	●	○ᵃ	●	○
dimension (dim)	○	○	●	○	●	○	○
dimension vector (vector)	○	○	●	○	●	○	○
system of units (system)	○	○	●	○	●	○	○
phenomenon (phen)	○	●	●	○	○	○	○
measurement (meas)	●	●	●	○	○	○	○
conversion (conv)	○	●	●	●	●	●	○
prefix (prefix)	●	○	●	●	●	●	●
unit ↔ system	○	○	●	○	●	○	○
unit ↔ qk	●	●	●	●	●	●	●
unit ↔ dim	○	○	●	○	●	○	○
unit ↔ vector	○	○	●	○	●	○	○
unit ↔ prefix	○	○	●	●	○	●	○
unit ↔ app	○	○	●	○	○	○	○
qk ↔ app	○	○	●	●	○	●	○
meas ↔ phen	○	●	●	○	○	○	○
meas ↔ qk	○	●	●	○	○	○	○
meas ↔ unit	●	●	●	○	○	○	○
meas ↔ value	●	●	●	○	○	○	○
symbols for units	●	○	●	●	●	●	○ᵃ
symbols for qks	○	○	●	○	●	○	○
typ. values per units and apps	○	○	○	○	○	○	○
typ. values per qks and apps	○	○	○	○	○	○	○
allowed values per units	○	○	○	○	○	○	○
allowed values per qks	○	○	○	○	○	○	○
precision of conversion	○	○	○	○	○	○	○
number of diff. lang. tagsᵇ	1	1ᶜ	3	0	0	0ᶜ	0
unit composition	○	○	●	○	○	○	○
quantity composition	○	○	○	○	○	○	○
everyday lang. designators	○	○	●	○	○	○	○
resolvable URIs	○ᵈ	○ᵈ	●	○	○	○ᵈ	●

ᵃInformation is included in the ontology, but not explicitly modeled using a specific concept or relation.
ᵇMissing language tags in labels result in a zero rating here.
ᶜLabels were almost always missing.
ᵈURIs did not resolve to concept specific websites, but to the whole ontology instead.

Table 3. Suitability scores for the examined ontologies.

		MUO	OBOE	OM	QU	QUDT	SWEET	UO
Group 1	UC 1[a]	0.27	0.20	0.71	0.40	0.47	0.40	0.20
	UC 2[a]	0.25	0.19	0.73	0.38	0.44	0.38	0.19
	UC 3[a]	0	0	0.89	0	0	0	0
	UC 4	0	1.00	1.00	0	0	0	0
Group 2	UC 5	0	1.00	1.00	1.00	1.00	1.00	0
	UC 6	0	0	0	0	0	0	0
Group 3	UC 7	0	0	1.00	0	1.00	0	0
	UC 8	0	0	1.00	0	0	0	0
	UC 9	0	0	0	0	0	0	0
	UC 10	1.00	1.00	1.00	1.00	1.00	1.00	1.00
	UC 11	0	0	0	0	0	0	0
Group 4	UC 12	0.27	0.27	1.00	0.36	0.82	0.36	0.27
	UC 13	0.75	0	1.00	0.75	1.00	0.75	0
	UC 14	0	0	1.00	0	0	0	0
	UC 15[a]	0.31	0.19	0.98	0.44	0.69	0.44	0.25
	UC 16[a]	0.29	0.29	0.95	0.43	0.57	0.43	0.43

[a]The result should only be used to compare ontologies, not to rate a single one.

Consistency checks (Group 3) just succeed for connections between unit of measurement and kind of quantity (UC 10). Other checks fail for different reasons with just a few exceptions: OM (UCs 7 and 8) and QUDT (UC 7).

Finally, the use of the ontology as a knowledge base (Group 4) seems pretty well supported. The only exception here is unit resolving (UC 14), which fails in all ontologies but OM due to the missing unit composition.

Overall there are just three use cases, that are currently not supported by any ontology. For each of those use cases, one crucial feature is missing:

- UC 6: Precision of Conversions.
- UC 9: Quantity Consistency.
- UC 11: Value Consistency.

From the point of view of a new project, OM seems to be the best choice right now. For no use case, any other ontology surpasses OM with respect to the suitability scoring with the closest overall contenders being QUDT, QU and SWEET.

6   Conclusion

We compiled an inventory of possible use cases for unit ontologies, grouped by similarity. This list consists of use cases given in literature as well as some, that have not been covered so far. We analyzed necessary as well as optional requirements. This resulted in the definition of a metric to compare the suitability of different ontologies for specific use cases. Using both requirement list and metric we then evaluated a set of seven representative ontologies.

The comparison highlighted the different focus in the development of the ontologies. Each one was created with a different set of use cases in mind. Summing up, current ontologies support a lot of use cases to a pretty decent level. However, our analysis reveals missing support for some use cases by ontologies.

Acknowledgments. Part of this work was funded by DFG in the scope of the LakeBase project within the Scientific Library Services and Information Systems (LIS) program.

References

1. Madin, J., Bowers, S., Schildhauer, M., et al.: An ontology for describing and synthesizing ecological observation data. Ecol. Inform. **2**(3), 279–296 (2007). doi:10.1016/j.ecoinf.2007.05.004
2. Rijgersberg, H., van Assem, M., Top, J.: Ontology of units of measure and related concepts. Semant. Web **4**(1), 3–13 (2013). doi:10.3233/SW-2012-0069
3. Hodgson, R., Keller, P.J., Hodges, J., Spivak, J.: QUDT - quantities, units, dimensions and data types ontologies, 18 March 2014. http://www.qudt.org/. Accessed 17 Aug 2016
4. Raskin, R.: Semantic Web for Earth and Environmental Terminology (SWEET). In: Earth Science Technology Conference. University of Maryland Inn., June 2003
5. Gkoutos, G.V., Schofield, P.N., Hoehndorf, R.: The Units Ontology: a tool for integrating units of measurement in science. In: Database (2012). doi:10.1093/database/bas033
6. Joint Committee for Guides in Metrology: International vocabulary of metrology. Basic and general concepts and associated terms. JCGM 200: 2012 (JCGM 200: 2008 with minor corrections), 3rd ed. (2012)
7. Foster, M.P.: Quantities, units and computing. Comput. Stan. Interfaces **35**(5), 529–535 (2013). doi:10.1016/j.csi.2013.02.001
8. Rijgersberg, H., Wigham, M., Top, J.: How semantics can improve engineering processes: a case of units of measure and quantities. Adv. Eng. Inform. **25**(2), 276–287 (2011). doi:10.1016/j.aei.2010.07.008. Information mining and retrieval in design
9. OASIS Quantities and Units of Measure Ontology Standard (QUOMOS) TC: Charter. https://www.oasis-open.org/committees/quomos/charter.php. Accessed 13 Sept 2016
10. Uschold, M., Gruninger, M.: Ontologies: principles, methods and applications. Knowl. Eng. Rev. **11**, 93–136 (1996). doi:10.1017/S0269888900007797
11. Hüsemann, B., Vossen, G.: Ontology engineering from a database perspective. In: Proceedings of the 10th Asian Computing Science Conference on Advances in Computer Science - ASIAN 2005, Data Management on the Web, Kunming, China, 7–9 December 2005, pp. 49–63 (2005). doi:10.1007/11596370_6
12. Tartir, S., Budak Arpinar, I.: Ontology evaluation and ranking using OntoQA. In: International Conference on Semantic Computing, ICSC 2007, pp. 185–192, September 2007. doi:10.1109/ICSC.2007.19.

13. Tello, A.L., Gómez-Péerez, A.: ONTOMETRIC: a method to choose the appropriate ontology. J. Database Manag. **15**(2), 1–18 (2004). doi:10.4018/jdm.2004040101
14. Resource Description Framework (RDF): Concepts and Abstract Syntax. Recommendation. W3C, 10 February 2004
15. Schadow, G., McDonald, C.J.: The unified code for units of measure. http://unitsofmeasure.org/ucum.html. Accessed 26 Apr 2016
16. United Nations Economic Commission for Europe (UNECE): Recommendation No. 20: Codes for Units of Measure Used in International Trade. Recommendation. UN/CEFACT Information Content Management Group (2009)
17. Schindler, S., Keil, J.M.: Unit ontology review v1.2.0, 8 September 2016. doi:10.5281/zenodo.61789
18. Schindler, S., Keil, J.M.: Unit ontology review results v1.2.0, 8 September 2016. doi:10.5281/zenodo.61787

A Simplified Agile Methodology
for Ontology Development

Silvio Peroni[(✉)]

DASPLab, DISI, University of Bologna, Bologna, Italy
`silvio.peroni@unibo.it`

Abstract. In this paper we introduce *SAMOD*, a.k.a. *Simplified Agile Methodology for Ontology Development*, a novel agile methodology for the development of ontologies by means of small steps of an iterative workflow that focuses on creating well-developed and documented models starting from exemplar domain descriptions. In addition, we discuss the results of an experiment where we asked nine people (with no or limited expertise in Semantic Web technologies and Ontology Engineering) to use SAMOD for developing a small ontology.

Keywords: Agile ontology development methodology · Conceptual modelling · Knowledge engineering · OWL Ontologies · Ontology engineering · SAMOD · Test-driven development

1 Introduction

Developing ontologies is not a straightforward task. This assumption is implicitly demonstrated by the number of ontology development processes that have been developed in last 30 years, that have their roots in the Knowledge and Software Engineering domains. Moreover, the choice of the right development process to follow is a delicate task, since it may vary according to a large amount of variables, such as the intrinsic complexity of domain to be modelled, the context in which the model will be used (enterprise, social community, high-profile academic/industrial project, private needs, etc.), the amount of time available for the development, and the technological hostility and the feeling of unfruitfulness shown by the final customers against both the model developed and the process adopted for the development.

In the past twenty years, the Software Engineering domain has seen the proposal of new *agile* methodologies for software development, in contrast with *highly-disciplined* processes that have characterised such discipline since its beginning. Following this trend, recently, agile development methodologies have been proposed in the field of Ontology Engineering as well (e.g. [3,7,13]). Such kind of methodologies would be preferred when the ontology to develop should be composed by a limited amount of ontological entities – while the use of

RASH: https://w3id.org/people/essepuntato/papers/samod-owled2016.html.

M. Dragoni et al. (Eds.): OWLED-ORE 2016, LNCS 10161, pp. 55–69, 2017.
DOI: 10.1007/978-3-319-54627-8_5

highly-structured and strongly-founded methodologies remain valid and, maybe, mandatory to solve and model incredibly complex enterprise projects.

One of main characteristics that ontology development methodologies usually have is the use of *exemplar data* during the development process so as to:

– *avoid inconsistencies* – a common mistake when developing a model is to make the TBox consistent if considered alone, and inconsistent when we define an ABox for it, even if all the classes and properties are completely satisfiable. Using real-world data, as exemplar of a particular scenario of the domain we are modelling, can definitely prevent this problem;
– *have self-explanatory and easy-understandable models* – trying to implement a particular real-world and significative scenario related to a model by using real data allows one to better understand if each TBox entity has a meaningful name that describes clearly the intent and the usage of the entity itself. This allows users to understand a model without spending a lot of effort in reading entity comments and the related documentation. The use of real data as part of the ontology development obliges ontology engineers and developers to think about the possible ways users will understand and use the ontology they are developing, in particular the very first time they look at it;
– *provide examples of usage* – producing data within the development process means to have a bunch of exemplars that describe the usage of the model in real-world scenarios. This kind of documentation, implicitly, allows users to apply a learn-by-example approach [1] in understanding the model and during their *initial skill acquisition* phase.

As already mentioned, several methodologies already propose the use of data during the development. However, the current ontology engineering processes, that deal with the development of small-/medium-size ontologies, usually do not include other aspects that, according to our experience, are crucial for guaranteeing a correct and quick outcome. In particular, it would be important:

– to take advantages of existing agile methodologies from the Software Engineering domain, by considering important features such as adaptive planning, evolutionary development, early delivery, continuous improvement, and rapid and flexible response to change;
– not to oblige pair programming – from our personal experience, the development of small ontologies usually involves only one ontology engineer;
– to provide a precise definition of different kinds of tests that the ontology must pass at each stage of the development, and that can be used for documenting the ontology as well.

In order to address all the aforementioned desiderata, in this paper we introduce *SAMOD* (*Simplified Agile Methodology for Ontology Development*), a novel agile methodology for the development of ontologies, partially inspired to the Test-Driven Development process in Software Engineering [2] and to existing agile ontology development methodologies such as eXtreme Design (XD) [13]. In particular, SAMOD is organised in three simple steps within an iterative process that focuses on creating well-developed and documented models by using significative exemplars of data, so as to produce ontologies that are always ready-to-be-used and easily-understandable by humans (i.e. the possible customers) without spending a lot of effort.

SAMOD is the result of our dedication to the development of ontologies in the past six years. While the first draft of the methodology has been proposed in 2010 as starting point for the development of the Semantic Publishing and Referencing Ontologies[1] [10], it has been revised several times so as to come to the current version presented in this paper – which has been already used for developing several ontologies, such as the Vagueness Ontology[2], the F Entry Ontology[3], the OA Entry Ontology[4], and the Imperial Data Ontology[5]. While a full introduction to SAMOD is provided in [11], in this paper we provide a summary of it and we discuss some outcomes of an user-based evaluation we have conducted in the past months.

The rest of the paper is organised as follows. In Sect. 2 we introduce the entities involved in the methodology. In Sect. 3 we present all the steps of SAMOD, providing details for each of them. In Sect. 4 we discuss the outcomes of an experiment where we asked to subjects with limited knowledge about Semantic Web technologies and Ontology Engineering to use SAMOD for developing an ontology. In Sect. 5 we present some of the most relevant related works in the area. Finally, in Sect. 6 we conclude the paper sketching out some future works.

2   Preliminaries

The kinds of people involved in SAMOD are domain experts and ontology engineers. A *domain expert*, or *DE*, is a professional with expertise in the domain to be described by the ontology, and she is mainly responsible to define, often in natural language, a detailed description of the domain in consideration. An *ontology engineer*, or *OE*, is a person who constructs meaningful and useful ontologies by using a particular formal language (such as OWL 2[6]) starting from an informal and precise description of a particular problem or domain provided by DEs.

A *motivating scenario* (MS) [17] is a small story problem that provides a short description and a set of informal and intuitive examples about it. In SAMOD, a motivation scenario is composed by a *name* that characterises it, a natural language *description* that presents a problem to address, and one or more *examples* according to the description.

An *informal competency question* (CQ) [17] is a natural language question that represents an informal requirement within a particular domain. In SAMOD,

---

[1] http://www.sparontologies.net/.

[2] http://www.essepuntato.it/2013/10/vagueness.

[3] http://www.essepuntato.it/2014/03/fentry.

[4] http://purl.org/emmedi/oaentry.

[5] http://www.essepuntato.it/2015/07/ido.

[6] http://www.w3.org/TR/owl2-syntax/.

each informal competency question is composed by an unique *identifier*, a natural language *question*, the kind of *outcome* expected as answer, some *exemplar answers* considering the examples provided in the related motivating scenario[7], and a list of identifiers referring to higher-level informal competency questions that the question in consideration *requires*, if any.

A *glossary of terms* (GoT) [5] is a list of term-definition pairs related to terms that are commonly used for talking about the domain in consideration. The term in each pair may be composed by one or more words or verbs, or even by a brief sentence, while the related definition is a natural language explanation of the meaning of such term. The terminology used for naming terms and for describing them must be as close as possible to the domain language.

As anticipated in the introduction, SAMOD prescribes an iterative process which aims at building the final model through a series of small steps. At the end of each iteration a particular preliminary version of the final model is released. Within a particular iteration i_n, the *current model* is the version of the final model released at the end of the iteration i_{n-1}. Contrarily, a *modelet* is a stand-alone model describing a particular aspect of the domain in consideration which is used to provide a first conceptualisation of a motivating scenario, without caring about the current model available after the previous iteration of the process – it is similar to a *microtheory* as introduced in Cyc [15]. By definition, a modelet does not include entities from other models and it is not included in other models.

A *test case* T_n, produced in the nth iteration of the process, is a sextuple including a motivating scenario MS_n, a list of scenario-related informal competency questions CQ_n, a glossary of terms GoT_n for the domain addressed by the motivating scenario, a $TBox_n$ of the ontology implementing the description introduced in the motivating scenario, an exemplar $ABox_n$ implementing all the examples described in the motivating scenario according to the $TBox_n$, and a set of SPARQL[8] queries SQ_n formalising the informal competency questions. A *bag of test cases (BoT)* is a set of test cases.

Given as input MS_n, $TBox_n$ and GoT_n – a *model test* aims at checking the validity of $TBox_n$ against specific requirements:

– **[formal requirement]** understanding (even by using appropriate unit tests [19]) whether $TBox_n$ is consistent;
– **[rhetorical requirement]** understanding whether $TBox_n$ covers MS_n and whether the vocabulary used by $TBox_n$ is appropriate.

Given as input MS_n, $TBox_n$ and $ABox_n$ built according to $TBox_n$, and considering the examples described in MS_n, a *data test* aims at checking the validity of the model and the dataset and against specific requirements:

---

[7] Note that if there are no data in any example of the motivating scenario that answer to the question, it is possible that either the competency question is not relevant for the motivating scenario or the motivating scenario misses some important exemplar data. In those cases one should remove the competency question or modify the motivating scenario accordingly.

[8] http://www.w3.org/TR/sparql11-query/.

- [**formal requirement**] understanding whether the $TBox_n$ is still consistent when considering the $ABox_n$;
- [**rhetorical requirement**] understanding whether the $ABox_n$ describes all the examples accompanying the motivating scenario completely.

Given as input $TBox_n$, $ABox_n$, CQ_n, and SQ_n, a *query test* aims at checking the validity of $TBox_n$, $ABox_n$, and each query in SQ_n against specific requirements:

- [**formal requirement**] understanding whether each query in SQ_n is well-formed and can correctly run on $Tbox_n + ABox_n$;
- [**rhetorical requirement**] understanding whether each query in CQ_n is mapped into an appropriate query in SQ_n and whether, running each of them on $TBox_n + ABox_n$, the result conforms to the expected outcome detailed in each query in CQ_n.

3   Methodology

SAMOD is based on the following three iterative steps (briefly summarised in Fig. 1) – where each step ends with the release of a snapshot of the current state of the process called *milestone*:

1. OEs collect all the information about a specific domain, with the help of DEs, in order to build a modelet formalising the domain in consideration, following certain ontology development principles. Then OEs create a new test case that includes the modelet. If everything works fine (i.e. model test, data test, and query test are passed), OEs release a milestone and proceed;

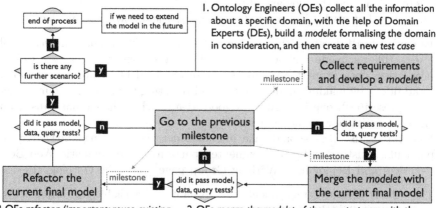

Fig. 1. A brief summary of SAMOD, starting with the "Collect requirements and develop a modelet" step.

2. OEs merge the modelet of the new test case with the current model produced by the end of the last iteration of the process, and consequently they update all the test cases in BoT specifying the new current model as *TBox*. If everything works fine (i.e. model, data and query tests are passed according to their formal requirements only), OEs release a milestone and proceed;
3. OEs refactor the current model, in particular focussing on the last part added in the previous step, taking into account good practices for ontology development processes. If everything works fine (i.e. model, data and query tests are passed), OEs release a milestone. In case there is another motivating scenario to be addressed, OEs iterate the process, otherwise the process ends.

The next sections elaborate on these steps introducing a real running example[9] considering a generic iteration i_n.

3.1 Step 1: Define a New Test Case

OEs and DEs work together to write down a motivating scenario MS_n, being as close as possible to the language DEs commonly use for talking about the domain. An example of motivating scenario is illustrated in Table 1.

Given a motivating scenario, OEs and DEs should produce a set of informal competency questions CQ_n, each of them identified appropriately. An example of an informal competency question, formulated starting from the motivating scenario in Table 1, is illustrated in Table 2.

Now, having both a motivating scenario and a list of informal competency questions, OEs and DEs write down a glossary of terms GoT_n. An example of glossary of terms is illustrated in Table 3.

The remaining part of this step is led by OEs only[10], who are responsible of developing a modelet according to the motivating scenario, the informal competency questions and the glossary of terms[11].

In doing that work, they must strictly follow the following principles:

- **Keep it small.** Keeping the number of the developed ontology entities small – e.g. Miller's magic number "7 ± 2" [9] entities per type (classes, object properties, data properties) – so as not to overload OEs' working memory. In addition, by making small changes (and retesting frequently, as our framework prescribes), one has always a good idea of what change has caused an error/inconsistency in the model [2].
- **Use patterns.** OEs should take into consideration existing knowledge, in particular existing and well-documented patterns – the Semantic Web Best Practices and Deployment Working Group page[12] and the Ontology Design

[9] The whole documentation about the example is available at http://www.essepuntato.it/2013/10/vagueness/samod.
[10] The OEs involved in our methodology can vary in number. However SAMOD has been thought for being used also by one OE only.
[11] Note that it is possible that multiple entities (i.e. classes, properties, individuals) are actually hidden behind one single definition in the glossary of terms.
[12] http://www.w3.org/2001/sw/BestPractices/OEP/.

Table 1. An example of motivating scenario.

Name	Vagueness of the TBox entities of an ontology
Description	Vagueness is a common human knowledge and language phenomenon, typically manifested by terms and concepts like High, Expert, Bad, Near etc. In an OWL ontology vagueness may appear in the definitions of classes, properties, datatypes and individuals. For these entities a more explicit description of the nature and characteristics of their vagueness/non-vagueness is required.
	Analysing and describing the nature of vagueness/non-vagueness in ontological entities is subjective activity, since it is often a personal interpretation of someone (a person or, more generally, an agent).
	Vagueness can be described according to at least two complementary types referring to quantitative or qualitative connotations respectively. The quantitative aspect of vagueness concerns the (real or apparent) lack of precise boundaries defining an entity along one or more specific dimensions. The qualitative aspect of vagueness concerns the identification of such other discriminants of which boundaries are not quantifiable in any precise way.
	Either a vagueness description, that specifies always a type, or a non-vagueness description provides at least a justification (defined either as natural language text, an entity or a more complex logic formula, or any combination of them) that motivates a specific aspect of why an entity should be intended as vague/non-vague. Multiple justifications are possible for the same description.
	The annotation of an entity with information about its vagueness is a particular act of tagging done by someone (i.e., an agent) who associates a description of vagueness/non-vagueness (called the body of the annotation) to the entity in consideration (called the target of the annotation).
Example 1	Silvio Peroni thinks that the class TallPerson is vague since there is no way to define a crisp height threshold that may separate tall from non-tall people.
	Panos Alexopoulos, on the other hand, considers someone as tall when his/her height is at least 190 cm. Thus, for Panos, the class TallPerson is not vague.
Example 2	In an company ontology, the class StrategicClient is considered vague. However, the company's R&D Director believes that for a client to be classified as strategic, the amount of its R&D budget should be the only factor to be considered. Thus according to him/her the vague class StrategicClient has quantitative vagueness and the dimension is the amount of R&D budget.
	On the other hand, the Operations Manager believes that a client is strategic when he has a long-term commitment to the company. In other words, the vague class StrategicClient has quantitative vagueness and the dimension is the duration of the contract.
	Finally, the company's CEO thinks that StrategicClient is vague from a qualitative point of view. In particular, although there are several criteria one may consider necessary for being expert (e.g. a long-standing relation, high project budgets, etc.), it's not possible to determine which of these are sufficient

Table 2. An example of competency question.

Identifier	3
Question	What are all the entities that are characterised by a specific vagueness type?
Outcome	The list of all the pairs of entity and vagueness type.
Example	StrategicClient, quantitative
	StrategicClient, qualitative
Depends on	1

Patterns portal[13] are both valuable examples – as well as widely-adopted Semantic Web vocabularies – such as FOAF[14] for people, SIOC[15] for social communities, and so on.

– **Middle-out development.** OEs should start to define the most relevant concepts and then to focus on more high-level and more concrete ones. Such middle-out approach [18] allows one to avoid unnecessary effort during the development because detail arises only as necessary, by adding sub- and super-classes to the basic concepts. Moreover, this approach, if used properly, tends to produce much more stable ontologies [17].

– **Keep it simple.** The modelet must be designed according to the information obtained previously (MS_n, CQ_n, GoT_n) in an as-quick-as-possible way, spending the minimum effort and without adding any unnecessary semantic structure – avoiding to think about inferences at this stage, and rather focussing on describing the motivating scenario fully.

– **Self-explanatory entities.** Each ontological entity must be understandable by humans by simply looking at its local name (i.e. the last part of the entity IRI). No labels and comments have to be added at this stage and all the entity IRIs must not be opaque – class local names has to be capitalised (e.g. *Justification*) and in camel-case notation if composed by more than one word (e.g. *DescriptionOfVagueness*), property local names must start with a non-capitalised verb[16] and in camel-case notation if composed by more than one word (e.g. *wasAttributedTo*), and individual local names must be non-capitalised (e.g. *ceo*) and dash-separated if composed by more than one word (e.g. *quantitative-vagueness*).

The goal of OEs is to develop a modelet$_n$, possibly starting from a graphical representation written in a proper visual language – such as Graffoo [4] – so as to convert it automatically in OWL by means of appropriate tools, e.g. DiTTO [6].

---

[13] http://www.ontologydesignpatterns.org/.
[14] http://xmlns.com/foaf/spec.
[15] http://rdfs.org/sioc/spec.
[16] http://www.jenitennison.com/blog/node/128.

Table 3. An example of glossary of terms.

Term	Definition
annotation of vagueness/non-vagueness	The annotation of an ontological entity with information about its vagueness is a particular act of tagging done by someone (i.e., an agent) who associates a description of vagueness/non-vagueness (called the body of the annotation) to the entity in consideration (called the target of the annotation).
agent	The agent who tags an ontology entity with a vagueness/non-vagueness description.
description of non-vagueness	The descriptive characterisation of non-vagueness to associate to an ontological entity by means of an annotation. It provides at least one justification for considering the target ontological entity non-vague. This description is primarily meant to be used for entities that would typically be considered vague but which, for some reason, in the particular ontology are not.
description of vagueness	The descriptive characterisation of vagueness to associate to an ontological entity by means of an annotation. It specifies a vagueness type and provides at least one justification for considering the target ontological entity vague.
vagueness type	A particular kind of vagueness that characterises the entity.
quantitative vagueness	A vagueness type that concerns the (real or apparent) lack of precise boundaries defining an entity along one or more specific dimensions.
qualitative vagueness	A vagueness type that concerns the identification of such other discriminants of which boundaries are not quantifiable in any precise way.
justification for vagueness/non-vagueness description	A justification that explains one possible reason behind a vagueness/non-vagueness description. It is defined either as natural language text, an entity, a more complex logic formula, or any combination of them.
has natural language text	The natural language text defining the body of a justification.
has entity	The entity defining the body of a justification.
has logic formula	The logic formula defining the body of a justification

Starting from $modelet_n$, OEs proceed in four phases:

1. run a model test on $modelet_n$. If it succeeds, then
2. create an exemplar dataset $ABox_n$ that formalises all the examples introduced in MS_n according to $modelet_n$. Then, OEs run a data test and, if succeeds, then

3. write SPARQL queries in SQ_n as many informal competency questions in CQ_n. Then, OEs run a query test and, if it succeeds, then
4. create a new test case $T_n = (MS_n, CQ_n, GoT_n, modelet_n, ABox_n, SQ_n)$ and add it to BoT.

When running the model test, the data test and the query test, it is possible to use any appropriate software to support the task, such as reasoners (Pellet[17], HermiT[18]) and query engines (Jena[19], Sesame[20]).

Any failure of any test that is considered a serious issue by the OEs results in getting back to the more recent milestone. It is worth mentioning that an exception should be also arisen if OEs think that the motivating scenario MS_n is to big to be covered by only one iteration of the process. In this case, it may be necessary to re-schedule the whole iteration, e.g. by splitting adequately the motivating scenario in two new ones.

3.2   Step 2: Merge the Current Model with the Modelet

At this stage, OEs merge $modelet_n$, included in the new test case T_n, with the current model, i.e. the version of the final model released at the end of the previous iteration (i.e. i_{n-1}). OEs proceed in three consecutive steps:

1. to define a new $TBox_n$ merging[21] the current model with $modelet_n$, by adding all the axioms in the current model and $modelet_n$ to $TBox_n$ and then by collapsing semantically-identical entities, e.g. those that have similar names and that represent the same real-world entity (for instance *Person* and *Human-Being*);
2. to update all the test cases in BoT, swapping the *TBox* of each test case with $TBox_n$ and refactoring each *ABox* and *SQ* according to the new entity names if needed, so as to refer to the more recent model;
3. to run the model test, the data test and the query test on all the test cases in BoT, according to their formal requirements only;
4. to set $TBox_n$ as the new current model.

Any serious failure of any test – i.e. something went bad in updating the test cases in BoT – results in getting back to a previous milestone. In this case, OEs have to consider either the most recent milestone, if they think there was a mistake in some actions performed during the current step, or one of the other previous milestones, if the failure is demonstrably a consequence of any of the components of the latest test case T_n.

---

[17] http://clarkparsia.com/pellet.
[18] http://hermit-reasoner.com/.
[19] http://jena.sourceforge.net/.
[20] http://www.openrdf.org/.
[21] If i_n is actually i_1, then the $modelet_n$ becomes the current model since no previous model is actually available.

3.3   Step 3: Refactor the Current Model

In the last step, OEs work to refactor the current model shared among all the test cases in BoT, and, accordingly, each *ABox* and *SQ* of each test case, if needed. In doing that task, OEs must strictly follow the following principles:

– **Reuse existing knowledge.** Reusing concepts and relations defined in other models is encouraged and often labelled as a common good practice [18]. The reuse can result either in including external entities in the current model as they are or in providing an *alignment*[22] or an *harmonisation*[23] with another model.
– **Document it.** Adding annotations – i.e. labels (i.e. *rdfs:label*), comments (i.e. *rdfs:comment*), and provenance information (i.e. *rdfs:isDefinedBy*) – to ontological entities, so as to provide natural language descriptions of them and to allow tools (e.g. LODE [12]) to produce an HTML human-readable documentation from the ontology source;
– **Take advantages from technologies.** Enriching the current model by using all the capabilities offered by OWL 2 (e.g. keys, property characteristics, property chains, inverse properties and the like) in order to infer automatically as much information as possible starting from a (possible) small set of real data. In particular, it is important to avoid over-classifications by specifying assertions that may be automatically inferred by a reasoner – e.g. creating an inverse property of a property P defining explicitly its domain and range even if they can be inferred automatically.

Finally, once the refactor is finished, OEs have to run the model test, the data test and the query test on all the test cases in BoT. This is a crucial task to perform, since it guarantees that the refactoring has not damaged any existing conceptualisation implemented in the current model.

3.4   Output of an Iteration

Each iteration of SAMOD produces a new test case that will be added to the bag of test cases (BoT). Each test case describes a particular aspect of the model under-development, i.e. the *current model* under consideration after one iteration of the methodology.

In addition of being integral part of the methodology process, each test case represents a complete documentation of a particular aspect of the domain described by the model, due to the natural language descriptions it includes (the motivating scenario, the informal competency questions, and the glossary of terms), as well as the formal implementation of exemplar data (the ABox) and possible ways of querying the data compliant with the model (the set of

---

[22] An alignment is set of correspondences between entities belonging to two models different models.

[23] It is the process of modifying a model (and also to align it, if necessary) to fully fit or include it into another model.

formal queries). All these additional information should help end-users in understanding, with less effort, what the model is about and how they can use it to describe the particular domain it addresses.

4   Experiment

We performed an experiment so as to understand to which degree SAMOD can be used by people with limited experience in Semantic Web technologies and Ontology Engineering. In particular, we organised a user testing session so as to gather some evidences on the usability of SAMOD when modelling OWL ontologies.

We asked nine Computer Science and Law people – one professor, two postdocs, and six Ph.D. students – to use SAMOD (one iteration only) for modelling a particular motivating scenario provided as exercise. SAMOD, as well as the main basics on Ontology Engineering, OWL, and Semantic Web technologies, were introduced to the subjects during four lectures of four hours each. At the end of the last lecture, we asked them to answer three questionnaires:

– a background questionnaire containing questions on previous experience in Ontology Engineering and OWL;
– another questionnaire containing ten likert questions according to the System Usability Scale (SUS), which also allowed us to measure the sub-scales of pure Usability and pure Learnability, as proposed recently by Lewis and Sauro [8];
– a final questionnaire asking for the experience of using SAMOD for completing the task.

The mean SUS score for SAMOD was 67.25 (in a 0 to 100 range), approaching the target score of 68 to demonstrate a good level of usability (according to [14]). The mean values for the SUS sub-scales Usability and Learnability were 65.62 and 73.75 respectively. In addition, an Experience score was calculated for each subject by considering the values of the answers given to the background questionnaire. We compared this score (x-axis in Fig. 2) with the SUS values and the other sub-scales (y-axis) using the Pearson's r. As highlighted by the red dashed lines (referring to the related Least Squares Regression Lines), there is a positive correlation between the Experience score and the SUS values – i.e. the more a subject knew about ontology engineering in general, the more SAMOD was perceived as usable and easy to learn. However, only the relation between the Learnability score and the Experience score was statistical significant ($p < 0.05$).

Axial coding of the personal comments expressed in the final questionnaires [16] revealed a small number of widely perceived issues. Overall the methodology proposed has been evaluated positively by 7 subjects (described with adjectives such as "useful", "natural", "effective", and "consistent"), but it has also received criticisms by 5 subjects, mainly referring to the need of more expertise in Semantic Web technologies and Ontology Engineering for using it appropriately. The use of the tests for assessing the ontology developed after a certain step has been

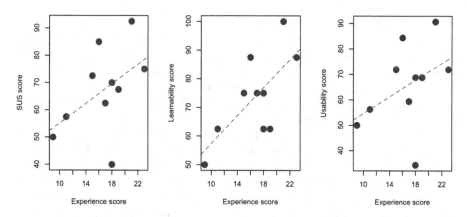

Fig. 2. Three comparisons between the SUS score (and its sub-scales) and the experience score by the subjects.

appreciated (3 positive comments vs. 1 negative one), as well as the use of the scenarios and examples in the very first step of SAMOD (3 positive comments) and the implementation of competency questions in form of SPARQL queries (2 positive comments). All the outcomes of the questionnaires are available online in the SAMOD GitHub repository[24].

5   Related Works

Several quick-and-iterative ontology development processes have been introduced recently, which could be preferred when the ontology to develop should be composed by a limited amount of ontological entities – while the use of highly-structured and strongly-founded methodologies (e.g. [5,17,18]) is still necessary and, maybe, mandatory for incredibly complex enterprise projects. In this section we introduce some of the most interesting agile approaches to ontology development.

One of the first agile methodologies introduced in the domain is *eXtreme Design (XD)* [13], which has been inspired by the eXtreme Programming methodology in Software Engineering. The authors described XD as "an approach, a family of methods and associated tools, based on the application, exploitation, and definition of ontology design patterns (ODPs) for solving ontology development issues". Summarising, XD is an agile methodology that uses pair design (i.e. groups of two ontology engineers working together during the development) and an iterative process which starts with the collection of *stories* and *competency questions* as requirements to address, and then it proposes the re-use of existing ontology design patterns for addressing such informal requirements.

Another recent approach has been introduced by Keet and Lawrynowicz in [7]. They propose to transfer concepts related to the Test-Driven Development in Software Engineering [2] into the Ontology Engineering world. The main idea

---

[24] http://github.com/essepuntato/samod.

behind this methodology is that tests have to be run in advance before to proceed with the modelling of a particular (aspect of a) domain. Of course, the first execution of the tests should fail, since no ontology has been already developed for addressing them properly, while the ontology developed in future iterations of the process should result in passing the test eventually.

De Nicola and Missikoff [3] have recently introduced their *Unified Process for ONtology building* methodology (a.k.a. *UPON Lite*), which is an agile ontology engineering method that places end-users without specific ontology expertise (domain experts, stakeholders, etc.) at the centre of the process. The methodology is composed by an ordered set of six steps. Each step outputs a self-contained artefact immediately available to end users, that is used as input of the subsequent step. This makes the whole process progressive and differential, and involves ontology engineers only the very last step of the process, i.e. when the ontology has to be formalised in some standard language.

6   Conclusions

In this paper we have introduced *SAMOD*, a *Simple Agile Methodology for Ontology Development*. In particular, we have introduced its process by detailing each of its steps, and we have also discussed the results of an experiment we have run involving nine people with no or limited expertise in Semantic Web technologies and Ontology Engineering.

In the future, we plan to involve a larger set of users so as to gather additional data about its usefulness, usability, and effectiveness. In addition, we plan to develop supporting tools for accompanying and facilitating users in each step of the methodology.

Acknowledgements. We would like to thank Jun Zhao for her precious comments and concerns about the initial drafts of SAMOD, David Shotton for our fruitful discussions when developing the SPAR Ontologies, Francesca Toni as one of the first users of such methodology, and Panos Alexopoulos as co-author of the Vagueness Ontology that we used herein to introduce all the examples of the SAMOD development process.

References

1. Atkinson, R.K., Derry, S.J., Renkl, A., Wortham, D.: Instructional principles from the worked examples research. Rev. Educ. Res. **70**(2), 181–214 (2000). http://dx.doi.org/10.3102/00346543070002181
2. Beck, K.: Test-Driven Development by Example. Addison-Wesley (2003). ISBN: 978-0321146533
3. De Nicola, A., Missikoff, M.: A lightweight methodology for rapid ontology engineering. Commun. ACM **59**(3), 79–86 (2016). http://dx.doi.org/10.1145/2818359
4. Falco, R., Gangemi, A., Peroni, S., Shotton, D., Vitali, F.: Modelling OWL ontologies with Graffoo. In: Presutti, V., Blomqvist, E., Troncy, R., Sack, H., Papadakis, I., Tordai, A. (eds.) ESWC 2014. LNCS, vol. 8798, pp. 320–325. Springer, Heidelberg (2014). doi:10.1007/978-3-319-11955-7_42

5. Fernandez, M., Gomez-Perez, A., Juristo, N.: METHONTOLOGY: from ontological art towards ontological engineering. In: Proceedings of the AAAI97 Spring Symposium Series on Ontological Engineering, pp. 33–40. http://aaaipress.org/Papers/Symposia/Spring/1997/SS-97-06/SS97-06-005.pdf

6. Gangemi, A., Peroni, S.: DiTTO: diagrams transformation inTo OWL. In: Proceedings of the ISWC 2013 Posters & Demonstrations Track (2013). http://ceur-ws.org/Vol-1035/iswc2013_demo_2.pdf

7. Keet, C.M., Ławrynowicz, A.: Test-driven development of ontologies. In: Sack, H., Blomqvist, E., d'Aquin, M., Ghidini, C., Ponzetto, S.P., Lange, C. (eds.) ESWC 2016. LNCS, vol. 9678, pp. 642–657. Springer, Heidelberg (2016). doi:10.1007/978-3-319-34129-3_39

8. Lewis, J.R., Sauro, J.: The factor structure of the system usability scale. In: Kurosu, M. (ed.) HCD 2009. LNCS, vol. 5619, pp. 94–103. Springer, Heidelberg (2009). doi:10.1007/978-3-642-02806-9_12

9. Miller, G.A.: Some limits on our capacity for processing information. Psychol. Rev. **63**(2), 81–97 (1956). http://dx.doi.org/10.1037/h0043158

10. Peroni, S.: The semantic publishing and referencing ontologies. In: Semantic Web Technologies and Legal Scholarly Publishing, pp. 121–193 (2014). http://dx.doi.org/10.1007/978-3-319-04777-5_5

11. Peroni, S.: SAMOD: an agile methodology for the development of ontologies. figshare (2016). http://dx.doi.org/10.6084/m9.figshare.3189769

12. Peroni, S., Shotton, D., Vitali, F.: The Live OWL Documentation Environment: a tool for the automatic generation of ontology documentation. In: Proceedings of EKAW 2012, pp. 398–412 (2012). http://dx.doi.org/10.1007/978-3-642-33876-2_35

13. Presutti, V., Daga, E., Gangemi, A., Blomqvist, E.: eXtreme design with content ontology design patterns. In: Proceedings of WOP 2009 (2009). http://ceur-ws.org/Vol-516/pap21.pdf

14. Sauro, J.: Background, Benchmarks & Best Practices (2011). ISBN: 978-1461062707

15. Sowa, J.F.: Representation and inference in the cyc project: D.B. Lenat and R.V. Guha. Artif. Intell. **61**(1), 95–104 (1993). http://dx.doi.org/10.1016/0004-3702(93)90096-T

16. Strauss, A., Corbin, J.: Basics of Qualitative Research Techniques and Procedures for Developing Grounded Theory, 2nd edn. Sage Publications, London (1998). ISBN 978-0803959408

17. Uschold, M., Gruninger, M.: Principles, methods and applications. IEEE Intell. Syst. **11**(2), 93–155 (1996). http://dx.doi.org/10.1109/MIS.2002.999223

18. Uschold, M., King, M.: Towards a methodology for building ontologies. In: Workshop on Basic Ontological Issues in Knowledge Sharing (1995). http://www.aiai.ed.ac.uk/publications/documents/1995/95-ont-ijcai95-ont-method.pdf

19. Vrandečić, D., Gangemi, A.: Unit tests for ontologies. In: Meersman, R., Tari, Z., Herrero, P. (eds.) OTM 2006. LNCS, vol. 4278, pp. 1012–1020. Springer, Heidelberg (2006). doi:10.1007/11915072_2

Using Ontology Design Patterns to Represent Sustainability Indicator Sets

Lida Ghahremanlou[1]([✉]), Liam Magee[2]([✉]), and James A. Thom[3]

[1] Coventry University, Coventry, UK
lida.ghahremanlou@bcu.ac.uk
[2] University of Western Sydney, Sydney, Australia
l.magee@westernsydney.edu.au
[3] RMIT University, Melbourne, Australia
james.thom@rmit.edu.au

Abstract. Sustainability indicators are increasingly being used to measure the economic, environmental and social properties of complex systems across different temporal and spatial scales. This motivates their inclusion in open distributed knowledge systems such as the Semantic Web. The diversity of such indicator sets provides considerable choice but also poses problems for those who need to measure and report. To address the modelling problems of indicator sets, we propose the use of Value Partition pattern to construct two design candidates: *generic* and *specific*. The generic design is more abstract, with fewer classes and properties, than the specific design. Documents describing two indicator systems – the Global Reporting Initiative and the Organisation for Economic Co-operation and Development – are used in the design of both candidate ontologies. We show the use of existing structural ontology design patterns can help to solve problems of ontology representations for modelling sustainability indicator sets.

Keywords: Sustainability Indicator Sets · Sustainability Reporting · Ontology Design Patterns · Value Partition

1 Introduction

Sustainability indicators estimate the past, current and future states of complex systems, such as cities, organisations, community groups and natural habitats. In a measurement context, a "system" is the entity that is the focus of various tasks that include identifying properties, devising scales, testing and measuring, and reporting on progress towards defined sustainability goals. In response to the demands of measuring and maintaining sustainability for diverse systems, many indicator sets have been developed and are in use today [4,16,19].

The diversity of such indicator sets provides considerable choice but also poses problems for those who need to measure and report. Often, relevant indicators need to be selected from multiple sets, with any gaps in specific measurement goals filled by the development of new indicators. Ontologies provide one

© Springer International Publishing AG 2017
M. Dragoni et al. (Eds.): OWLED-ORE 2016, LNCS 10161, pp. 70–81, 2017.
DOI: 10.1007/978-3-319-54627-8_6

means for consolidating these multiple sets in a single representation, but leave open the problem of exactly how this representation is designed. In many cases, it remains an advantage for such a representation to be human-readable as well as machine-readable. This facilitates interpretation of how different sets compare and contrast. To support human usability in the sustainability domain, any such representation should aim to support the easy *reading* of existing indicators compiled from heterogeneous sources, and the easy *writing* of new indicators and annotations to the ontology through common authoring tools such as *Protégé Desktop*[1].

We argue *ontology design patterns* can help to address both problems in a way that is systematic and builds upon the experience of others. Our focus in this paper is on the first of these problems: **How to represent indicators from multiple indicator sets in an ontology?** This problem includes a further semantic challenge, since multiple sets may overlap at the level of individual concepts but may also overlap between broader conceptual clusters. We argue this challenge in turn has at least two levels: (i) indicators may be *named differently*, due to different languages, disciplinary jargon, or designer preferences; and (ii) indicators may also be *conceptually organised differently*, due to the knowledge paradigms and priorities motivating indicator selection. In both cases, merging two or more indicator sets into a single, combined ontology can assist in identifying which specific indicators might be most relevant to the measurement task at hand.

Well-known standardised frameworks for sustainability reporting include the Global Reporting Initiative (GRI) indicators and guidelines[2], the Organization for Economic Co-operation and Development (OECD)[3] and the United Nations Statistics Division (UN Social Indicators)[4]. Each of these frameworks group sustainability indicators into hierarchical structures that include categories and sub-categories of indicators. Extracts of GRI and OECD indicator sets are shown in Fig. 1, which illustrates (i) categories (or aspects), (ii) sub-categories (themes) and (iii) indicators. This shows, at least at a structural level, that there is some basis for comparison between these two widely used sets of sustainability indicators.

To date, there have been few efforts to represent *multiple* sustainability indicator sets in a systematic semantic way. Advantages of representing indicators in a formal *ontology* include developing a consistent definition of what an indicator is, how it can be applied, and how it relates to higher order grouping constructs used in theories and definitions of sustainability. An ontology representation also builds upon the many tools now available for ontology reasoning, alignment and visualisation, allowing organisations to browse and review different kinds of indicators for different measurement applications. Most importantly, by utilising pre-defined matches between non-identical but related indicators, measurements

---

[1] http://protege.stanford.edu/.

[2] http://www.globalreporting.org/.

[3] http://www.oecd.org/.

[4] http://www.un.org/esa.

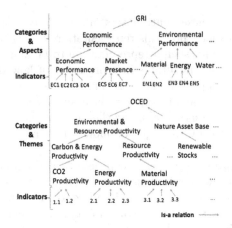

Fig. 1. Extracts of GRI and OECD indicator sets

and reports developed by different organisations and contexts can be more easily compared.

A key concern in ontology engineering is to design and organise groups of related concepts that capture the relevant information of the domain being modelled as an ontology. *Ontology Design Patterns* (ODPs) have been proposed to encourage compatibility, efficiency and recognisability of ontology designs [14,17]. In the formal sense provided by those listed[5], patterns make explicit relations that would otherwise remain implicit, or at best only documented. As one example, the *Role* pattern[6] makes clear that two ontology classes are not simply related through user-defined properties, but are related specifically as *task actions* and *role objects*.

In this paper we discuss two ontology design candidates, which we term *generic* and *specific*, developed to represent sustainability indicator sets. We have termed the target end ontology OSIS (Ontology for Sustainability Indicator Sets), and the two design candidates GOSIS and SOSIS. The details of ontology engineering steps are described in earlier work [13]. This paper instead discusses the varied use of an ODP called *Value Partition* in the construction of the two candidates, and presents conclusions on the relevant merits of each variation.

2   Related Work

To prepare our discussion of the two ontologies, we review briefly literature relating to (i) sustainability indicator sets and (ii) ontology design patterns.

---

[5] http://ontologydesignpatterns.org.
[6] http://ontologydesignpatterns.org/wiki/Submissions:Role_task.

2.1 Ontologies and Taxonomies Used in Sustainability Indicator Sets

There have been several attempts to develop domain and application ontologies in the context of sustainability and sustainability reporting. Brilhante et al. [4] present an ontology that aims to represent economic indicators of sustainable development. Similarly, Madlberger et al. [18] develop an ontology for the domain of corporate sustainability, heavily influenced by the design of the Global Reporting Initiative's *XBRL* specification. Kumazawa et al. [16] outline an ontological approach to capture a very broad problem-based definition of sustainability science, developed around five key concepts of *Problem, Goal, Evaluation, Countermeasure* and *Domain Concept.* Han and Stoffel [15] apply text extraction and analysis techniques to environmental sustainability case studies to generate machine and humanly-readable ontologies. An ontology-based approach has also been used by Pinheiro et al. [19] to assist selection of relevant sustainability indicators. Finally, Fox [7] has developed an ontology to represent ISO37120 Global City Indicators, a standard that defines measures for urban sustainability.

This prior work has not sought to combine more than one representation of sustainability indicators into a single ontology design. To help address this problem, we next examine ontology design patterns.

2.2 Ontology Design Patterns

Ontology design patterns borrow heavily from the related concept of *Software Design Patterns* (SDPs) [8] in software engineering. Using object oriented SDPs provides software class models with well-understood properties and behaviours that solve common engineering challenges in generic, abstract and reusable ways. As a result, such patterns improve software development efficiency and generate high-quality and more maintainable software artefacts [3]. In an equivalent way, an ontology can be composed of different related ODPs, which resemble building blocks that make up the ontology structure. Recognising generic or abstract ontology components is an integral part of specifying appropriate ODPs. This process is often domain-dependent, and thus requires deep understanding of the key concepts of the domain problem. Similar to SDPs, ODPs are abstract, flexible and reusable solutions that address common problems and use cases in the field of ontology engineering [1,2]. However, given that ontology engineering is a less mature field compared with software engineering, the definition, representation and application of ODPs lack the same level of consensus as software engineering design patterns.

The ODP literature can be divided into studies that discuss general ideas about ODPs and those that discuss concrete ODPs for tackling specific design problems in developing ontologies. As examples of the former, Reich [21] first introduced the notion of ODPs in the context of molecular biology. Shortly after, Staab et al. [23] discussed the idea of Semantic Patterns and Knowledge Patterns as reusable components for building knowledge bases. Their work was later

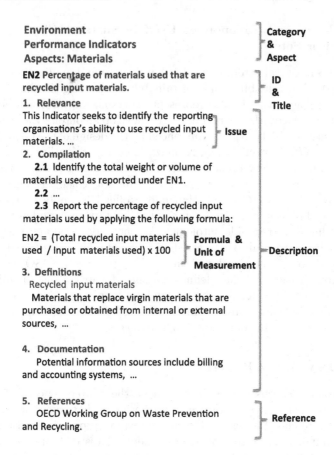

Fig. 2. An extract of a GRI indicator

followed by the work of Gangemi [9] and Gangemi et al. [11] that distinguished between Logical, Conceptual and Content Ontology Design Patterns. Finally, Gangemi and Presutti [10] classify a number of commonly-used ontology patterns into six major categories including: Structural, Correspondence, Content, Reasoning, Presentation, and Lexico-Syntactic ODPs. This classification system continues to influence their organisation on the widely-used pattern repository[7].

---

[7] http://ontologydesignpatterns.org.

3   Sustainability Indicator Sets Ontology Development

In this section we discuss the development process of our ontology design.

3.1   Extracting Foundational Concepts

After reviewing GRI and OECD indicator sets, and interviewing sustainability domain experts, we first identify several common foundational concepts of sustainability indicators. Highlighted in Fig. 2, these concepts include: *Indicator, IndicatorSet, Category, SubCategory (Group, Theme, Aspect), Issue, Description (Relevance, Compilation, Definitions, Documentation), Reference (Sources, Information).*

3.2   Modelling Problems

Second, given the identified key concepts within the domain-dependent taxonomies, we further identify relations between these concepts and the relevant entities within those taxonomies. These may have quite different representations. In particular, we have noticed that specific GRI and OECD indicator systems can be specified in relation to abstract concepts of `IndicatorSet` and `Indicator` in different ways. The question here is of **how to determine whether such relations be represented as disjoint class hierarchies, as subclasses of a common parent class, or as instances of a given class**, and represents a refinement of the overall research question of how to represent indicators from multiple sets in an ontology. In concrete terms, as we discuss in the next subsection, the design problem involves the association of the `Indicator` concept with the `IndicatorSet` concept. This also affects the relations of other concepts such as `Category`, `Description` and `Reference`. Addressing these modelling problems ideally should reflect the requirements of the final ontology design, leading us to choose appropriate patterns that satisfy both computational properties and the human interpretation of ontologies. These ontology requirements for sustainability indicator sets are discussed in our earlier work [12].

3.3   Ontology Design Patterns Solution

Third, to address the aforementioned modelling problems, we decided upon the use of the Value Partition (VP) pattern. The VP ontology pattern was first introduced by Rector [20] and further reviewed and developed by Aranguren [1] for the biomedical domain. The VP pattern represents specified collections of "values" – also known as a "feature space" – using hierarchical modelling. Generally speaking, in any domain, such characteristics are used to describe different concepts in the ontology. For example, given the description of the "IndicatorSet" concept in the sustainability domain, in the presented ontology model, there are two VP patterns as follows.

As Rector [20] notes, the VP pattern can be implemented in different ways: as a collection of individuals, as disjoint classes, or as datatypes. As the values we are modelling form themselves complex taxonomic structures, datatypes are not adequate. Accordingly we present two approaches to OSIS that represent multiple indicator sets, respectively, as collections of interrelated individuals and as disjoint classes. In doing so, we also acknowledge subsequent clarifications of complex uses of the Value Partition ODP in, for example, Rodriguez-Castro et al. [22], who relate VP to two other ODPs: Normalisation and Class as a Property Value (or CPV). In line with their findings, both of our VP implementations simultaneously constitute implementations of Normalisation and what they term strict-CPV ODPs.

- **Pattern 1 - GOSIS design:** This design assumes indicator sets and indicators are instances of classes. Both GRI and OECD taxonomic structures are represented as individuals of the `IndicatorSet` class. Specific measures, such as GRI's "Percentage of total employees covered by collective bargaining agreements" (Disclosure 102-41) and OECD's "share of the population connected to sewerage with primary, secondary, tertiary treatment", are represented as instances of the `Indicator` class. The `Indicator` class is further linked by a particular property, `belongsToIndicatorSet`, to the `IndicatorSet` class. This view is broader to cover sustainability indicators' key information with no reference to any particular organisations and is called *Generic Ontology for Sustainability Indicator Set* or GOSIS.

  In our discussions with domain experts, the particular affordance of this design is its *reusability* and *extensibility*. People who are not ontology designers can add new indicator sets and indicators without modifying the classes and properties of the ontology.
- **Pattern 2 - SOSIS design:** This design treats indicator sets and their indicators as disjoint class hierarchies. For example, `GRIIndicatorSet` is a subclass of the `IndicatorSet` superclass, while indicators are instances of classes that model properties specific to each indicator system. This allows for a more direct representation of the underlying conceptualisation of those systems. The `GRI Relevance` class has no equivalent concept in the OECD taxonomy, and this difference is evident in the class design of the ontology alone. Since this view includes direct references to specific indicator sets, it is called *Specific Ontology for Sustainability Indicator Sets* or SOSIS.

  Domain experts considered this design more helpful in terms of *explicitness*, as it is clear what information is available about indicators in each of the distinct indicator sets. However new indicator sets require additional modelling of the ontology's classes and properties, which impacts its *reusability*.

4 Discussion

Following the ontology engineering process of METHONTOLOGY [6] described in earlier works [12,13] and using the Value Partition pattern – described in the

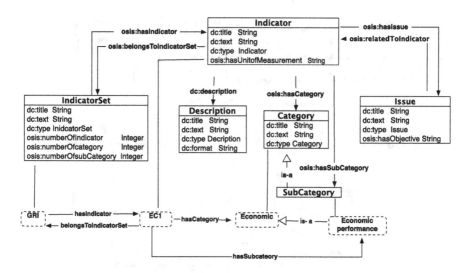

Fig. 3. UML diagram of GOSIS design using Value Partition pattern 1

series of WC3 practices[8] and discussed in Sect. 3.3 – we have developed two ontology design candidates, labelled respectively GOSIS and SOSIS. These differ largely in terms of abstraction, as discussed below.

The GOSIS design defines broadly a suitable structure and reflects the generic key concepts of sustainability indicators. As a result, and in line with pattern 1, this design applies an object-oriented approach that encapsulates the generic features of all indicator sets into a series of base or foundational classes. The SOSIS design, on the other hand, emphasises the role of the organisations that develop sustainability indicators. In designing SOSIS, we use VP pattern 2, that includes the key concepts of these organisations with their own indicator classifications. As a result, this design uses a range of classes and relations that are specifically added for each sustainability indicator set.

The UML diagrams, built upon the aforementioned Patterns of VP, are shown in Figs. 3 and 4 and the OWL representation of both ontology designs can be found here[9].

The GOSIS design treats each indicator set as well as their indicators as individuals, and each set instantiates properties and relations of the `IndicatorSet` class (Fig. 3). It contains fewer classes, and is less intuitive for domain experts to read – at least in an ontology editing tool such as *Protégé Desktop*, understanding the structure of the ontology requires frequent traversal of 'Class' and 'Individual' tabs, for instance. Accordingly, we consider this a more abstracted view of the underlying domain of multiple sustainability indicator systems.

By contrast, the SOSIS design treats each indicator set as a class. Accordingly, they inherit rather than instantiate properties and relations of the `IndicatorSet`

[8] http://www.w3.org/2001/sw/BestPractices/OEP/SpecifiedValues-20050223/.

[9] http://www.circlesofsustainability.org/wp-content/uploads/2016/12/.

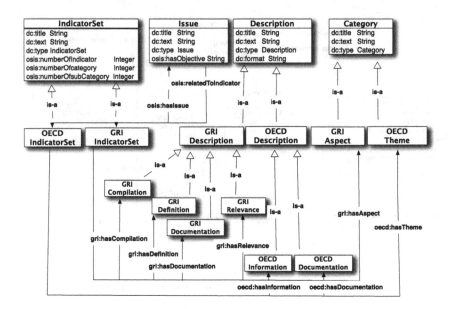

Fig. 4. UML diagram of SOSIS design using Value Partition pattern 2

class. This produces a much larger ontology that maps directly to the specific frames of reference that it is derived from, and we term this the *concrete* variation of the Value Partition ontology design pattern (Fig. 4).

5 Conclusion and Future Work

In this work, we have discussed how the use of existing ontology design patterns can help resolve modelling issues in developing and constructing an ontology for sustainability indicators.

Our focus in designing GOSIS and SOSIS was to employ the Value Partition ODP to develop generic and specific models for sustainability indicators that covers broad key concepts of the domain as well as specific indicator sets. The findings from the previous section indicate the relative merits of our ontology designs. From a human readability perspective, we determine the two candidates, the generic design GOSIS and the specific design SOSIS, differ largely in terms of their relative abstractness or concreteness. The generic design contains less classes, and is less readable; the specific design has more classes, but is more difficult to modify or extend.

We have previously evaluated these ontology design candidates in earlier work [13]. Based on the findings presented here, we conclude that the specific design is preferable where the domain requirements require a high degree of fidelity to existing and known frames of reference, while the generic design offers greater reuse in contexts where unseen and unknown sets of indicators need to be added to the ontology in an *ad hoc* fashion. Accordingly, we also suggest that both ontology design models have their distinct merits, satisfying different requirements for representing indicator systems. Such requirements are generality and reusability in the case of GOSIS, and precision and intuitiveness in the case of SOSIS.

Our conclusion is aligned with one of the limitations of the VP ontology design pattern, which was developed based on OWL 1 in 2005 [20]. The constraints with OWL 1 was that a class in ontology must not be the value of a property. According to Rodriguez-Castro et al. [22], this constraint is resolved in OWL 2, where a class can have a property or instance values at the same time for DL reasoner. In addition, recent studies [5,24] propose mapping structural design patterns in OWL as new solutions for such constraints.

Further work can be undertaken to incorporate additional sustainability indicators systems, and to further refine the candidate OSIS ontologies presented in this research. One approach for incorporating new systems is through the automation facilities provided by ontology matching algorithms[10]. Though discouraged by Rector [20], we also anticipate the possibility of blending the generic and specific designs in future, possibly using Simple Knowledge Organization System (SKOS)[11] as a means for representing complex sets of values that align to the IndicatorSet and Indicator classes of the generic and more abstract design. Recent work by Dudáš et al. [5] has also proposed PURO, a partially automated approach to generating alternative OWL encodings. Future work aims to examine how the respective merits of both designs can be preserved, and the ontology extended to other indicators systems, using combinations of ontology matching, SKOS representations and PURO software. Once complete, we also aim to conduct an axiom-based comparison of the two designs, to evaluate formally their respective similarities, differences and merits.

Acknowledgements. We thank the anonymous reviewers for their valuable comments. This research has been supported in part by an Australian Research Council (ARC) funded linkage project LP0990509 on *Accounting for Sustainability: Developing an Integrated Approach for Sustainability Assessments.*

References

1. Aranguren, M.E.: Ontology design patterns for the formalisation of biological ontologies. Master's thesis, University of Manchester (2005)
2. Aranguren, M.E.: Role and Application of Ontology Design Patterns in Bio-Ontologies. PhD thesis, University of Manchester (2009). http://mikeleganaaranguren.files.wordpress.com/2010/01/thesis.pdf
3. Booch, G.: Object-Oriented Analysis and Design with Applications. Benjamin-Cummings, Redwood City (1994)
4. Brilhante, V., Ferreira, A., Marinho, J., Pereira, J.S.: Information integration through ontology and metadata for sustainability analysis. In: The International Environmental Modelling and Software Society (iEMSs) 3rd Biennial Meeting (2006)

---

[10] http://ontologymatching.org/projects.html.
[11] https://www.w3.org/2004/02/skos/.

5. Dudáš, M., Hanzal, T., Svátek, V., Zamazal, O.: OBOWLMorph: Starting ontology development from PURO background models. In: Tamma, V., Dragoni, M., Gonçalves, R., Ławrynowicz, A. (eds.) OWLED 2015. LNCS, vol. 9557, pp. 14–20. Springer, Cham (2016). doi:10.1007/978-3-319-33245-1_2

6. Fernandez, M., Gomez-Perez, A., Juristo, N.: METHONTOLOGY: From ontological art towards ontological engineering. In: Proceedings of the AAAI97 Spring Symposium Series on Ontological Engineering, pp. 33–40. AAAI Press (1997)

7. Fox, M.S.: A foundation ontology for global city indicators. Department of Mechanical and Industrial Engineering University of Toronto, Global Cities Institute Working Paper No. 3 (2014)

8. Gamma, E., Helm, R., Johnson, R., Vlissides, J.: Design Patterns: Elements of Reusable Object-oriented Software. Pearson Education, Upper Saddle River (1994)

9. Gangemi, A.: Ontology design patterns for semantic web content. In: Gil, Y., Motta, E., Benjamins, V.R., Musen, M.A. (eds.) ISWC 2005. LNCS, vol. 3729, pp. 262–276. Springer, Heidelberg (2005). doi:10.1007/11574620_21

10. Gangemi, A., Presutti, V.: Ontology design patterns. In: Staab, S., Studer, R. (eds.) Handbook on Ontologies. International Handbooks on Information Systems, pp. 221–243. Springer, Heidelberg (2009). doi:10.1007/978-3-540-92673-3_10

11. Gangemi, A., Gómez-Pérez, A., Presutti, V., Suárez-Figueroa, M.C.: Towards a catalog of OWL-based ontology design patterns. In: CAEPIA. Neon Project Publications (2007)

12. Ghahremanloo, L.: An integrated knowledge base for sustainability indicators. In: Australasian Computing Doctoral Consortium. RMIT Melbourne (2012). http://www.cs.rmit.edu.au/acdc2012/

13. Ghahremanloo, L., Thom, J.A., Magee, L.: An ontology derived from heterogeneous sustainability indicator set documents. In: Proceedings of the Seventeenth Australasian Document Computing Symposium, pp. 72–79. ACM (2012)

14. Guarino, N., Welty, C.: Evaluating ontological decisions with OntoClean. Commun. ACM 45(2), 61–65 (2002). doi:10.1145/503124.503150

15. Han, D., Stoffel, K.: Ontology based qualitative case studies for sustainability research. In: Proceedings of the AI for an Intelligent Planet. ACM (2011). Article 6

16. Kumazawa, T., Saito, O., Kozaki, K., Matsui, T., Mizoguchi, R.: Toward knowledge structuring of sustainability science based on ontology engineering. Sustain. Sci. 4(1), 99–116 (2009). doi:10.1007/s11625-008-0063-z

17. Lozano-Tello, A., Gómez-Pérez, A.: ONTOMETRIC: a method to choose the appropriate ontology. J. Database Manage. 15(2), 1–18 (2004). http://oa.upm.es/6467/

18. Madlberger, L., Thöni, A., Wetz, P., Schatten, A., Tjoa, A.M.: Ontology-based data integration for corporate sustainability information systems. In: Proceedings of International Conference on Information Integration and Web-Based Applications & Services, pp. 353–357. ACM (2013)

19. Pinheiro, W.A., Barros, R., De Souza, J.M., Xexeo, G.B., Strauch, J., Barros, P., Campos, M.: Adaptative methodology of sustainability indicators management by ontology. Int. J. Glob. Environ. Issues 9(4), 338–355 (2009). http://EconPapers.repec.org/RePEc:ids:ijgenv:v:9:y:2009:i:4:p:338-355

20. Rector, A. (ed.) Representing specified values in OWL: "value partitions" and "value sets". W3C working group note, 17 May 2005

21. Reich, J.R.: Ontological design patterns: metadata of molecular biological ontologies, information and knowledge. In: Ibrahim, M., Küng, J., Revell, N. (eds.) DEXA 2000. LNCS, vol. 1873, pp. 698–709. Springer, Heidelberg (2000). doi:10.1007/3-540-44469-6_65
22. Rodriguez-Castro, B., Ge, M., Hepp, M.: Alignment of ontology design patterns: class as property value, value partition and normalisation. In: Meersman, R., et al. (eds.) OTM 2012. LNCS, vol. 7566, pp. 682–699. Springer, Heidelberg (2012). doi:10.1007/978-3-642-33615-7_16
23. Staab, S., Erdmann, M., Maedche, A.: Engineering ontologies using semantic patterns. In: Proceedings of the IJCAI 2001 Workshop on E-Business & the Intelligent Web, pp. 174–185 (2001)
24. Svátek, V., Homola, M., Kluka, J., Vacura, M.: Mapping structural design patterns in owl to ontological background models. In: Proceedings of the Seventh International Conference on Knowledge Capture, pp. 117–120. ACM (2013)

Application of Inference Rules to a Software Requirements Ontology to Generate Software Test Cases

Vladimir Tarasov[1(✉)], He Tan[1], Muhammad Ismail[1], Anders Adlemo[1], and Mats Johansson[2]

[1] School of Engineering, Jönköping University, Box 1026, 551 11 Jönköping, Sweden
{vladimir.tarasov,he.tan,muhammad.ismail,anders.adlemo}@ju.se
[2] Saab AB, Slottsgatan 40, 551 11 Jönköping, Sweden
mats.e.johansson@saabgroup.com

Abstract. Testing of a software system is resource-consuming activity. One of the promising ways to improve the efficiency of the software testing process is to use ontologies for testing. This paper presents an approach to test case generation based on the use of an ontology and inference rules. The ontology represents requirements from a software requirements specification, and additional knowledge about components of the software system under development. The inference rules describe strategies for deriving test cases from the ontology. The inference rules are constructed based on the examination of the existing test documentation and acquisition of knowledge from experienced software testers. The inference rules are implemented in Prolog and applied to the ontology that is translated from OWL functional-style syntax to Prolog syntax. The first experiments with the implementation showed that it was possible to generate test cases with the same level of detail as the existing, manually produced, test cases.

Keywords: Inference rules · Ontology · OWL · Prolog · Requirement specification · Test case generation

1 Introduction

In modern society software products and systems permeates every aspect of our lives, such as our homes, cars, the public infrastructure and even our bodies. As a consequence, quality concerns are becoming much more vital and critical as we get more dependent on these products and systems. The yearly cost of software errors as a consequence of poor quality procedures in the software industry was estimated to roughly $312 billion, according to a report in 2013 by the Cambridge University [8], and the cost still continues to increase. As detection of software errors goes hand-in-hand with testing, the same increase in cost is true for all kind of software testing activities [3,8].

One way of curbing this ongoing trend is to automate as many as possible of the software test activities. As far as test case execution goes, this is already

© Springer International Publishing AG 2017
M. Dragoni et al. (Eds.): OWLED-ORE 2016, LNCS 10161, pp. 82–94, 2017.
DOI: 10.1007/978-3-319-54627-8_7

a mature field where commercial products help software testers in their daily work. The automatic generation of test cases, however, is an entirely different matter.

The use of ontologies for testing has not been discussed as much as the use of ontologies in other stages of the software development process. In [6] the authors discussed possible ways of utilizing ontologies for the test case generation, and the feasibility of reuse of domain knowledge encoded in ontologies for testing. In practice, however, few results have been presented in the area. Most of them have had a focus on testing web-based software and especially web services (e.g. [14,17]).

This paper proposes an approach to combine an OWL ontology with inference rules in order to construct an ontology-based application. The purpose of the application is to generate software test cases based on a software requirements specification[1]. The ontology, which represents both the software requirements and the software, hardware and communication components belonging to an embedded system, is translated from OWL functional-style syntax into Prolog syntax. The inference rules, that represent the expertise of an expert software tester, are coded in Prolog and make use of the ontology entities to generate test cases. The Prolog inference engine controls the process of selecting and invoking inference rules.

The rest of this paper is structured as follows. Section 2 details the proposed approach, including the ontology representing software requirements and software components, inference rules capturing strategies for test case generation, and the OWL to Prolog translation. Section 3 presents an evaluation of the approach. Related work is described in Sect. 4. Our conclusions on the result are given in Sect. 5.

2   Approach to Test Case Generation

When testers create test cases, they do it based on software requirements specifications and their own expertise, expertise that comes from previous work on testing software systems. To automate this process, both parts should be represented in machine-processable form. The requirements are usually described in semi-structured text documents or stored in requirements management systems. This part is captured in a requirements specification ontology, which conceptualizes the structure of software requirements and their relations to different components of a software system (Sect. 2.1). The second part, the testers' expertise is less structured and is acquired by interviewing experienced testers and studying existing test cases with their corresponding requirements. Such knowledge is represented with inference rules that utilize the ontology for checking conditions and querying data (Sect. 2.3). To make it possible to use the ontology entities together with the inference rules, it is necessary to translate the ontology to a format supported by the rules (Sect. 2.2).

---

[1] The study presented in this paper is part of the project Ontology-based Software Test Case Generation (OSTAG).

2.1 Representation of Requirements with Requirements Specification Ontology

The ontology in this paper includes three pieces of knowledge: (1) a meta model of the software requirements, (2) the domain knowledge of the application, e.g. general knowledge of the hardware and software, the electronic communication standards etc., and (3) each system requirements specifications. The components that make up the domain ontology come from an embedded system within an avionic system provide by Saab Avionics. In this work the ontology is used to support test case generation. It can also be used to support other tasks in the software development process, such as requirement analysis, and requirement verification and validation.

The current version of the ontology contains 42 classes, 34 object properties, 13 datatype properties, and 147 individuals in total. Figure 1 presents the meta model of the software requirements. As indicated in the figure, each requirement is concerned with certain functionalities of the software component. For example, a requirement may be concerned with data transfer. Each requirement consists of at least (1) requirement parameters, which are inputs of a requirement, (2) requirement conditions, and (3) results, which are usually outputs of a requirement, and exception messages. Some requirements require the system to take actions. Furthermore, there exists traceability between different requirements, e.g. traceability between an interface requirement and a system requirement.

Figure 2 shows the ontology fragment for one particular functional requirement, SRSRS4YY-431. If the communication type is out of its valid range, the initialization service shall deactivate the UART (Universal Asynchronous Receiver/Transmitter), and return the result "comTypeCfgError". In Fig. 2, the rectangles represent the concepts of the ontology; the round rectangles represent the individuals; and the dashed rectangles provide the data values of datatype property for individuals. More details about the ontology can be found in [16].

2.2 OWL-to-Prolog Translation

To prepare the ontology for the use by the inference rules, it is necessary to translate it into the syntax that is supported by the rules. As soon as Prolog

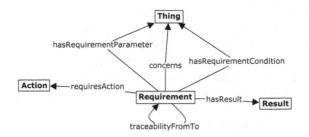

Fig. 1. The meta model of a requirement in the ontology

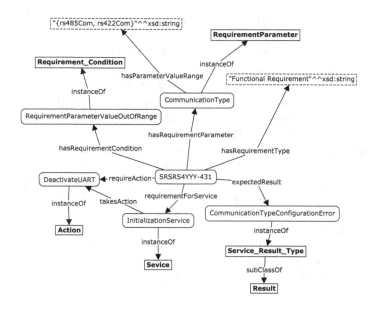

Fig. 2. Ontology fragment for the SRSRS4YY-431 requirement specification

is chosen for coding inference rules (see Sect. 2.3), the ontology constructs have to be translated into the Prolog syntax. There exist a number of serialisation formats that can be used to save an OWL ontology to a file: RDF/XML, Turtle, OWL/XML, Manchester OWL syntax or functional-style syntax. The functional-style syntax is the closest one to the Prolog syntax. An ontology document in the functional-style syntax is a sequence of OWL constructs enclosed in the Ontology statement as well as a number of prefix definitions [11]. As a logical consequence, we have chosen functional-style syntax as the starting point for the ontology translation.

A Prolog program consists of clauses. The term "clause" denotes a fact or a rule in a knowledge base. A clause is ended with full stop (.) and different terms in a clause are separated with commas (,). The basic terms that are used in Prolog programs are atoms, numbers, variable and structures [2]:

- An atom is a string of characters that starts with a lower-case letter,
- A variable is a string of characters that starts with an upper-case letter,
- Integers and real numbers (floating point numbers) are also allowed in Prolog,
- Structures or complex data objects are objects with several components. Functor is used to combine several components into a single one, e.g. "date(14, June, 2006)".

When an ontology is written in the functional-style syntax, every single line is a separate statement that represents one construct. Each line is processed separately to translate it into the corresponding Prolog statement. A Python script has been written for the OWL-to-Prolog translation, which performs these steps for every OWL statement:

Table 1. Example of translation of some OWL statements

OWL functional-style syntax	Prolog syntax
ClassAssertion(OSTAG:Error_Handling_ Requirement :SRSRS4YY-431)	classAssertion(error_handling_require- ment, srsrs4yy_431)
DataPropertyAssertion(:hasParameter- ValueList :NumberOfStopBits "[stopBits1, stop-Bits2]"^^xsd:string)	dataPropertyAssertion(hasParameter- ValueList, NumberOfStopBits, [stopBits1, stopBits2])
DataPropertyDomain(:hasParameter- ValueList :Requirement_Parameter)	dataPropertyDomain(hasParameter- ValueList, requirement_Parameter)
ObjectPropertyAssertion(OSTAG: requirementForService :SRSRS4YY-431 :InitializationService)	objectPropertyAssertion(require- mentForService, srsrs4yy_431, initializationService)
SubClassOf(OSTAG:Error_Handling_ Requirement OSTAG:Requirement)	subClassOf(error_handling_ requirement, requirement)
AnnotationAssertion(rdfs:label OSTAG:FIFO "FIFO")	annotationAssertion(rdfslabel, fifo, 'FIFO')

- Read an OWL statement and remove OWL prefixes[2],
- Tokenize the statement and convert each token into lowerCamelCase notation because Prolog atoms start with a lower case latter.
- Convert the list of tokens into a Prolog clause in the form of a fact.

The following OWL statements are translated at the moment: ClassAssertion, subClassOf, ObjectPropertyAssertion, DataPropertyAssertion, objectProperty-Range, objectPropertyDomain, annotationAssertion. Table 1 shows several examples of translation from OWL to Prolog.

2.3   Deriving Test Cases from the Ontology with Inference Rules

To derive test cases from the ontology, it is necessary to represent testers' exper-tise on how they use requirements to create test cases. This kind of knowledge is less structured and more difficult to capture. Few general guidelines can be found in literature, such as boundary value testing. However, most expertise is specific to particular types of software systems and/or particular domains. To capture this expertise or knowledge it is necessary to interview experienced testers and study existing test cases and their corresponding requirements. Such knowledge embodies inherent strategies for test case creation, knowledge that can be expressed in the form of heuristics represented as if-then rules.

In this study we examined 16 requirements and 20 corresponding test cases. Each requirement describes some functionality of a service (function) from a driver for a hardware unit, in this case represented by an embedded avionic

---

[2] There is only one ontology used at the moment but if there are imported ontologies in the future, prefixes can be translated as well.

system component. Thus, all requirements are grouped according to services. We analysed requirements covering six services. During the analysis an original test case, previously created manually by a software tester, was compared with the corresponding requirement to fully understand how different parts of the original test case had been constructed. Then, any inconsistencies or remaining doubts were cleared during discussions with the industry software testers participating in the study.

The outcome from these activities was a set of inference rules formulated in plain English. Each original test case consists of four parts: prerequisite conditions, test inputs, test procedure, and expected test results. Consequently, inference rules were formulated for each of the test case parts. An example of a inference rule for the test procedure part of the requirements SRSRS4YY-431 is shown below:

```
IF the requirement is for a service and a UART controller is to be
   deactivated
THEN add the call to the requirement's service, calls to a trans-
   mission service and reception service as well as a recovery
   call to the first service.
```

The condition (if-part) of a heuristic rule is formulated in terms of the individual representing the requirement and the related ontology entities representing connected hardware parts, input/output parameters for the service and the like. The action (then-part) part of the rule contains instructions on how a test case part is to be generated.

After formulating the inference rules, they need to be implemented in a programming language. There are two basics requirements that have to be met by such a language: (1) it should have means to represent the rule in a natural way and (2) it should have means to access the entities in the ontology. We chose Prolog [2] as the language for the implementation as Prolog complies with both of the basic requirements. The acquired inference rules can be implemented with the help of Prolog rules (a Prolog rule is analogous to a statement in other programming languages). After the OWL-to-Prolog translation (described in the previous sub-section), the ontology becomes an inherent part of the Prolog program, and, as a consequence, the ontology entities can be directly accessed by the Prolog code. Finally, the inference engine that is built-in into Prolog is used to execute the coded rules to generate test cases.

An example of the inference rule written in Prolog that implements the previous heuristic rule is given below:

```
1 tc_procedure(Requirement, Procedure) :-
     % get service individual for call #1
2    objectPropertyAssertion(requirementForService, Requirement,
        Service),
     % check condition for calls #2-4
```

```
3   objectPropertyAssertion(requiresAction, Requirement,
       DeactivateUART),
4   objectPropertyAssertion(actsOn, DeactivateUART,
       UartController),
5   classAssertion(uart_controller, UartController),
    % get individuals of the required services
6   classAssertion(transmission_service, WriteService),
7   classAssertion(reception_service, ReadService),
8   Procedure = [Service, WriteService, ReadService,
       recovery(Service)].
```

Line 1 in the example is the head of the rule consisting of the name, "input" argument and "output" argument. Lines 2–7 encode the condition of the heuristic as well as acts as queries to retrieve the relevant entities from the ontology. Line 8 constructs the procedure part of the test case as a list of terms. The list is constructed from the retrieved ontology entities and special term functors.

Figure 3 shows the ontology entities used by the inference rule, when it is invoked to generate a test case for the requirement SRSRS4YY-431. The figure shows ontology paths, each one being a number of ontology entities connected by object properties or subsumption relation.

Each test case is generated sequentially, from the prerequisites part through to the results part. The generated parts are collected into one structure by the following rule:

```
test_case(Requirement,
  tc(description(TCid, ReqID, Service), Prerequisites, Inputs,
     Procedure, Results)) :-
  req_id(Requirement, ReqID),
  objectPropertyAssertion(requirementForService, Requirement,
     Service),
  tc_prerequisites(Requirement, Prerequisites),
  tc_inputs(Requirement, Inputs),
  tc_procedure(Requirement, Procedure),
  tc_results(Requirement, Results),
  new_tcid(TCid).
```

Finally, the test case structure is translated into plain text in English. The final result can be found in the right column in Table 3.

3   Experiment and Evaluation

The example provided by Saab consisted of a hardware module with embedded code. The examined part of the example consisted of 15 requirements, specified in the SRS (Software Requirement Specification document), with corresponding 18 test cases, specified in the STD (Software Test Description document). In

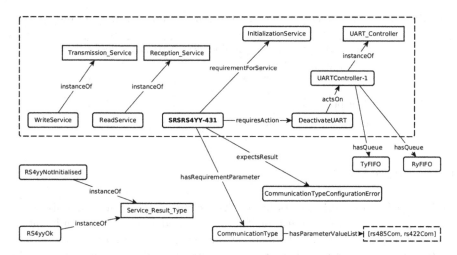

Fig. 3. Ontology paths used by the inference rules to generate a test case for the requirement SRSRS4YY-431. The dashed line indicates the paths used by the inference rule demonstrated in the example above to generate the test procedure part of the test case. The other paths are used by the remaining rules to generate the other parts of the test case.

most cases one requirement is evaluated by executing one test case but in some occasions one requirement is evaluated by executing two or more test cases.

A total of 40 inference rules were used to generate the 18 test cases. The number of rules for each test case part is detailed in Table 2 (an auxiliary rule is intended to be invoked by a main rule). The corresponding test cases have been reproduced in plain English, using the same format as described in the STD document, by applying the inference rules to the ontology. The result from this exercise can be observed in Table 3 where the text in left column is a slightly modified excerpt from the STD document while the text in the right column is the generated output through applying some of the inference rules to the ontology. The result presented in the table corresponds to one specific requirement, in this case SRSRS4YY-431, a requirement that is evaluated in one test case, in this case STDRS4YY-114. As can be observed, there is an almost one-to-one correspondence between the texts in the two columns. However, the authors would like to point out that in some occasions the generated test case texts indicated a discrepancy with the corresponding test case texts found in the STD document. These discrepancies were presented to and evaluated by personnel from Saab and on occasions the observed discrepancies indicated a detected error in the STD document. Hence, this correctness insurance exercise helped improving the quality of the STD document.

Table 2. Number of inference rules used to generate different parts of test cases

Test case part	Main rules	Auxiliary rules
Prerequisite conditions	8	4
Test inputs	8	1
Test procedure	5	
Expected test tesults	9	5

4   Related Work

The reason for conducting tests on a software product/system is mainly to be able to put some level of trust on the quality and requirement fulfilment of the product/system. To run the tests on the product/system, some kind of test case(s) must be designed and the corresponding test code(s) be programmed. In many occasions, if not most, this is a manual activity with everything that this embodies of potential errors in the test code caused by missed or misinterpreted requirements due to a deficient test case description or a non-experienced tester.

In an attempt to counteract on these negative effects, model-driven testing techniques have surged in recent years as an alternative field of applied research in the software testing domain [1,12]. One specific modelling language that has emerged as the prime modeling-tool in this domain is UML. There have been presented a large number of projects with a focus on automatic generation of test cases based on the usage of UML to describe some parts of the testing activities [9] Other examples of model-driven test case generation projects have been based on Function Block Diagrams [4] or State-based testing [7], just to mention two. The different model-driven test case generation approaches presented by different researcher teams often depend on some kind of requirement specification as input to the process [9]. When it comes to the focus of the test activities, i.e. what is the output from the testing activities that needs to be evaluated, two main areas can be identified, code coverage testing (which could be looked upon as testing the output of a software design process) and requirement coverage testing (which could be looked upon as testing the input to a software design process). All of the presented model-driven test case generation approaches referred to earlier have had a focus on some kind of code coverage. However, in some application domains the verification of the coverage of the requirements, which means that all requirements stated in a requirements specification document have been considered and tested in a traceable manner, is equally, and sometimes even more, important than code coverage. This is the case in, for example, the avionics industry of which Saab is a perfect example.

There exist only a few projects that rely on ontologies for software testing activities, for example [5,13]. As mentioned earlier in this paper, an ontology represents a formal model of the knowledge captured for a specific domain, in this paper being software testing. However, it should be stressed that the creation of an ontology is only the first step in order to automatically create software

Table 3. Test case from the STD (left column) and the corresponding generated test case by applying inference rules to the ontology (right column)

...	...
Test Inputs	**Test Inputs:**
1. According to table below.	1. \<communicationType\> := min_value - 1
2. \<uartId\> := \<uartId\> from the rs4yy_init call	\<communicationType\> := max_value + 1
3. \<uartId\> := \<uartId\> from the rs4yy_init call	\<communicationType\> := 485053
4. \<comType\> := rs4yy_rs422Com	2. \<uartID\> := \<uartID\> from the initializationService call
	3. \<uartID\> := \<uartID\> from the initializationService call
	4. \<communicationType\> := RS422
Test Procedure	**Test Procedure:**
1. Call rs4yy_init	1. Call initializationService
2. Call rs4yy_write	2. Call writeService
3. Call rs4yy_read	3. Call readService
4. Recovery: Call rs4yy_init	4. Recovery: Call initializationService
Expected Test Results	**Expected Test Results:**
1. \<result\> == rs4yy_comTypeCfgError	1. \<result\> == communicationTypeConfigurationError
2. \<result\> == rs4yy_notInitialised	2. \<result\> == rs4yyNotInitialised
3. \<result\> == rs4yy_notInitialised, \<length\> == 0	3. \<result\> == rs4yyNotInitialised, \<length\> == 0
4. \<result\> == rs4yy_ok	4. \<result\> == rs4yyOk
...	...

test cases. It must also be contemplated that the test cases must be generated with some specific test objectives in mind. The OSTAG-project that has been presented in this paper is one of very few examples where both code coverage and requirement coverage can be handled.

Prolog has been used as a reasoner for OWL ontologies in a number of cases. For example, in [15] the authors describe an approach to reasoning over temporal ontologies that translates OWL statements to clauses in Prolog and then uses the built-in inference mechanism. In [10] an OWL ontology and OWLRuleML rules are translated into Prolog clauses, which are then used to infer new facts by the Prolog inference engine. The work presented in this paper has utilised a similar idea, however, we have used OWL functional-style syntax for the OWL to Prolog translation, which makes queries to the ontology as close as possible to OWL syntax.

5   Conclusions

We have proposed an approach to generate software test cases based on the use
of an ontology, representing software requirements as well as knowledge about
the components of the software system under development, and inference rules,
representing strategies for test case creation. The inference rules are coded in
Prolog and the built-in inference engine is used for executing the rules. During
the execution the rules query the ontology to check conditions and retrieve data
needed for the construction of test cases. To make this possible, the ontology is
serialised in OWL functional-style syntax and then translated to Prolog syntax.
The first experiment showed that, by using 40 inference rules, 18 test cases for
15 requirements were generated as plain text in English. The examination of
the result showed an almost one-to-one correspondence between the texts in the
generated test cases and the texts provided by one of our industrial partners,
Saab.

The translation from the OWL functional-style syntax to the Prolog syntax
allowed for seamless integration of the ontology into the Prolog program. On
one hand, the syntax of the OWL statements was preserved to a great extent.
On the other hand, the inference rules could directly reference the ontology
constructs in their bodies. The Prolog inference mechanism took care of finding
an inference rule with a satisfied condition and firing it. As a result, the ontology
was effectively used for an applied purpose—automation of software testing.
However, it should be noted that not all OWL statements are translated at the
moment. Most notably complex class constructors are not translated (due to the
fact that we did not find the need to use them in the ontology so far). There
is also lack of inference rules preserving the semantics of OWL, e.g. rules to
find all individuals of a class having several subclasses. Moreover, the conducted
experiment is of limited scale. More experiments with an increased number of
inference rules are needed to evaluate the proposed approach to demonstrate its
full potential.

There exist other languages to implement inference rules, e.g. SWRL or the
inference rule language built-in in Jena, which are closer to the syntax and
semantics of OWL and follow the open world assumption. Such languages may
be better suited for situations when new data need to be integrated into the
knowledge base. Despite that, we have chosen Prolog because our case does not
require data integration. Additionally, Prolog provides both inference mechanism and traditional programming facilities, thus, eliminating the need to use
one language for implementing inference rules and another one for developing
a software prototype. At the same time, OWL was chosen as the language to
implement the ontology to support test case generation because the ontology
can also be used to support other tasks in the software development process,
such as requirement analysis, and requirement verification and validation.

The future work will go along the lines of increasing the number of inference
rules to generate test cases for the so far uncovered requirements. This will
allow us to further test the applicability of the proposed approach of combining
an OWL ontology and Prolog inference rules in an ontology-based application,

such as in the software test case generation domain. A comparison can also be done between the Prolog and OWL reasoning systems.

So far the results from the project have been positive and have demonstrated the feasibility of producing test cases in a semi-automatic fashion. The automation of the test case generation process has demonstrated that the correctness of the generated test cases was also improved. Minor errors that went undetected by the human test case designers were identified and corrected as mentioned in Sect. 3. This result puts the finger on the benefits of automating a process in general and the test case generation process in specific. However, this is not the only measurable result that is expected to come from the project. In the near future other types of metric are going to be evaluated, such as to quantify the time savings gained from automating the test case generation process through real-life time studies, and to verify the coverage of the requirements to demonstrate that all requirements stated in a requirements specification document have been considered and tested.

Acknowledgments. The research reported in this paper has been financed by grant #20140170 from the Knowledge Foundation (Sweden).

References

1. Anand, S., Burke, E., Chen, T., Clark, J., Cohen, M., Grieskamp, W., Harman, M., Harrold, M., McMinn, P.: An orchestrated survey on automated software test case generation. J. Syst. Softw. **86**(8), 1978–2001 (2013)
2. Bratko, I.: Prolog Programming for Artificail Intelligence, 4th edn. Pearson Education, Upper Saddle River (2011)
3. CapGemini, HP, Sogeti: World quality report 2015–16, 80 p. (2015)
4. Enoiu, E., Causevic, A., Ostrand, T., Weyuker, E., Sundmark, D., Pettersson, P.: Automated test generation using model-checking: an industrial evaluation. Int. J. Softw. Tools Technol. Transf. **1**(1), 1–19 (2014)
5. Freitas, A., Vieira, R.: An ontology for guiding performance testing. In: 2014 IEEE/WIC/ACM International Joint Conferences on Web Intelligence (WI) and Intelligent Agent Technologies (IAT), pp. 400–407 (2014)
6. Happel, H.J., Seedorf, S.: Applications of ontologies in software engineering. In: Proceedings of Workshop on Sematic Web Enabled Software Engineering (SWESE) on the ISWC, pp. 5–9 (2006)
7. Holt, N., Briand, L., Torkar, R.: Empirical evaluations on the cost-effectiveness of state-based testing: an industrial case study. Inf. Softw. Technol. **56**, 890–910 (2014)
8. Judge Business School, Cambridge University: Cambridge university study states software bugs cost economy \$312 billion per year (2013). http://www.prweb.com/releases/2013/1/prweb10298185.htm. Accessed 22 Sept 2016
9. Kaur, A., Vig, V.: Systematic review of automatic test case generation by UML diagrams. Int. J. Eng. Res. Technol. (IJERT) **1**(6), 17 (2012)
10. Laera, L., Tamma, V., Bench-Capon, T., Semeraro, G.: SweetProlog: a system to integrate ontologies and rules. In: Antoniou, G., Boley, H. (eds.) RuleML 2004. LNCS, vol. 3323, pp. 188–193. Springer, Heidelberg (2004). doi:10.1007/978-3-540-30504-0_15

11. Motik, B., Patel-Schneider, P., Parsia, B.: OWL 2 Web Ontology Language: Structural Specification and Functional-Style Syntax. W3C, 2nd edn. (2012)

12. Mussa, M., Ouchani, S., Al Sammane, W., Hamou-Lhadj, A.: A survey of model-driven testing techniques. In: QSIC 2009 9th International Conference on Quality Software, 24–25 August 2009, Jeju, South Korea, pp. 167–172 (2009)

13. Nasser, V.H., Du, W., MacIsaac, D.: Knowledge-based software test generation. In: The 21st International Conference on Software Engineering and Knowledge Engineering, Boston, USA, pp. 312–317, July 2009

14. Nguyen, C.D., Perini, A., Tonella, P.: Ontology-based test generation for multiagent systems. In: Proceedings of the 7th international Joint Conference on Autonomous Agents and Multiagent Systems, vol. 3, pp. 1315–1320 (2008)

15. Papadakis, N., Stravoskoufos, K., Baratis, E., Petrakis, E., Plexousakis, D.: PROTON: a prolog reasoner for temporal ontologies in OWL. Expert Syst. Appl. **38**(12), 14660–14667 (2011)

16. Tan, H., Muhammad, I., Tarasov, V., Adlemo, A., Johansson, M.: Development and evaluation of a software requirements ontology. In: 7th International Workshop on Software Knowledge-SKY 2016 in Conjunction with the 9th International Joint Conference on Knowledge Discovery, Knowledge Engineering and Knowledge Management-IC3K 2016 (2016)

17. Wang, Y., Bai, X., Li, J., Huang, R.: Ontology-based test case generation for testing web services. In: Eighth International Symposium on Autonomous Decentralized Systems, ISADS 2007, pp. 43–50. IEEE (2007)

Collaborative Ontology Evolution and Data Quality - An Empirical Analysis

Nandana Mihindukulasooriya[✉], María Poveda-Villalón, Raúl García-Castro, and Asunción Gómez-Pérez

Ontology Engineering Group, Universidad Politécnica de Madrid, Madrid, Spain
{nmihindu,mpoveda,rgarcia,asun}@fi.upm.es

Abstract. Since more than a decade, theoretical research on ontology evolution has been published in literature and several frameworks for managing ontology changes have been proposed. However, there are less studies that analyze widely used ontologies that were developed in a collaborative manner to understand community-driven ontology evolution in practice. In this paper, we perform an empirical analysis on how four well-known ontologies (*DBpedia, Schema.org, PROV-O,* and *FOAF*) have evolved through their lifetime and an analysis of the data quality issues caused by some of the ontology changes. To that end, the paper discusses the composition of the communities that developed the aforementioned ontologies and the ontology development process followed. Further, the paper analyses the changes in those ontologies in the 53 versions of them examined in this study. Depending of the use case, the community involved, and other factors different approaches for the ontology development and evolution process are used (e.g., bottom-up approach with high automation or top-down approach with a lot of manual curation). This paper concludes that one model for managing changes does not fit all. Furthermore, it is also clear that none of the selected ontologies follow the theoretical frameworks found in literature. Nevertheless, in communities where industrial participants are dominant more rigorous editorial processes are followed, largely influenced by software development tools and processes. Based on the analysis, the most common quality problems caused by ontology changes include the use of abandoned classes and properties in data and introduction of duplicate classes and properties.

Keywords: Ontology · Change · Evolution · Quality · DBpedia · Schema.org · PROV-O · FOAF

1 Introduction

Ontologies are defined as formal, explicit specifications of a shared conceptualization of a domain of interest [1] and inherently ontology development becomes a collaborative process due to the "shared conceptualization" aspect. Once an ontology is defined, changes to it are inevitable. The main causes of ontology

© Springer International Publishing AG 2017
M. Dragoni et al. (Eds.): OWLED-ORE 2016, LNCS 10161, pp. 95–114, 2017.
DOI: 10.1007/978-3-319-54627-8_8

change include (i) changes in the domain; (ii) changes in conceptualization; and (iii) changes in the explicit specification [2]. As the knowledge about the domain of interest evolves dynamically over time, ontology changes have to occur to reflect those changes in the conceptualization. Changes may occur not only because of the changes in the reality but also because of the decisions of the ontology engineer. The ontology engineer may decide to expand the scope of the conceptualization by adding new concepts and relations or to apply other design patterns and strategies requiring modifications to the existing ontology both in structure and in semantics, or to remove existing terms due to various reasons, for instance, because some existing terms become obsolete or cause confusion in users.

The elements that are subject to change in ontologies include concepts, properties, instances and axioms. These changes impact the entities that are described using these ontologies as well as any other ontologies that are dependent on the given ontology. In the ontology evolution topic, research looked into several problems and challenging tasks. These problems include consistency maintenance, backward compatibility, ontology manipulation, understanding of ontology evolution, change propagation, etc. In this paper, we focus on analyzing the changes in different versions of the ontology and on discussing their effect on data quality.

According to Stojanovic [3], ontology evolution refers to the activity of facilitating the modification of an ontology by preserving its consistency, while the ontology modification activity would consist on changing the ontology without considering the consistency. For the sake of clarity we will use the term "ontology evolution" regardless of whether consistency is preserved.

Furthermore, ontology engineers are not always aware of or have any centralized control over the ontology users. Thus, it is not always possible for them to notify the changes in an ontology to their users. And, even if those changes were notified, there could be data that use a previous version of the ontology.

Starting in the early nineties, theoretical research on ontology evolution has been published in literature [2,4] and frameworks for managing ontology evolution [5–7] have been developed. However, to the best of the authors' knowledge there are no studies that analyze how such research was used in ontologies developed in a collaborative manner. In this regard, this paper provides an empirical study on how four widely-used ontologies have evolved through their lifetime.

The rest of the paper is organized as follows. Section 2 presents an overview of pioneer works on ontology evolution and a review of the quality issues that ontology evolution might cause. Next, Sect. 3 describes the ontologies selected to be analyzed. Section 4 details the obtained results and the analysis carried out for each of the selected ontologies. Finally, Sect. 5 presents some conclusions and suggestions for next steps.

2   Ontology Evolution

First steps towards ontology evolution were proposed by Noy and Klein [2] in the early nineties. They presented an analysis of the traditional versioning and evolution dimensions that apply to databases and discussed whether they can be applied to ontologies. They also provided a comprehensive list of possible changes affecting the ontology and their impact on the instance data (mostly oriented to DAML-OIL ontologies).

From a methodological point of view, the DILIGENT [8] methodology has been one of the first approaches for supporting distributed ontology development in a distributed environment taking into account the evolution of ontologies. The general process proposed comprises five main activities: build, local adaptation, analysis, revision, and local update.

Regarding approaches based on OWL ontologies, we can mention the process proposed by Stojanovic [3] that enables handling ontology changes ensuring the consistency of the ontology. Along this work a taxonomy of ontology changes is proposed based on whether the changes are additive or subtractive applied to certain ontology elements or meta changes.

Finally, Palma [5] proposed an ontology metadata model to provide a high-level overview of how an ontology has changed. Such model is the basis for the methods proposed in his work for ontology change management in distributed environments and the strategies for supporting collaborative ontology development. In this work, not only a plain list of possible changes is provided but a more complex model that includes atomic changes, entity changes and composite changes.

More recently, Dragoni and Ghidini [9] explored ontology evolution from the information retrieval tools point of view. In this work, the authors analyze how changes in ontologies might affect information retrieval performance in terms of precision and recall. In this case the modifications made on an ontology are defined in terms of patterns that might occur over taxonomical information. The three patterns considered are "Rename" (change of the label used for identifying the concept), "Delete" (removal of a concept from the ontology) and "Move" (change of the position of the concept in the ontology).

The above-mentioned works established definitions and first steps towards ontology change managements and automation. However, to the best of the authors' knowledge there are less studies that analyze how such research was used in widely-used ontologies developed in a collaborative manner.

In this paper, we look at a subset of the change operations already proposed in the literature. First, we identify the main meta elements defined in the OWL 2 Web Ontology Language (RDF Semantics). In this study, we mainly focus on (a) `rdfs:Class`, (b) `owl:Class`, (c) `rdf:Property`, (d) `owl:ObjectProperty`, and (e) `owl:DatatypeProperty`.

2.1 Quality Issues Caused by Ontology Evolution

When parts in a given ontology change and evolve, they can cause impact on the entities described using them as well as on any other ontologies that are dependent on them [3].

On the one hand, when changes cause inconsistencies in the ontology itself, they can be identified through validation, for instance, using a reasoner to check for inconsistencies. However, some problems are not so evident and are hard to detect automatically. For instance, if a redundant property is added that has the same semantics of an existing property (same meaning but different term/identifier), it decreases the conciseness quality but it is not straightforward to detect it.

On the other hand, when the changes in the ontology cause problems in data instances described using the ontology as well as in other ontologies that import the given ontology, it is hard to detect those problems. Most often the ontology developers are not aware of all the downstream users of the ontology and at the moment there are no well-established methods for automatically notifying the users of an ontology about changes in new versions of it. Though there are several approaches in the literature [10,11], there is a lack of practical tools for determining the impact of changes of a given on ontology on data instances and dependent ontologies. If we take the analogy of software development, when a dependency changes in a software project compilation, errors, unit test and integration test failures reveal any potential problems due to changes in the dependency. Such practices are not yet well-established in the ontology engineering community.

In the literature, changes to the ontology are categorized into two main groups: structural changes and semantic changes [12]. On the one hand, structural changes include addition, deletion, renaming, splitting/merging a concept, property, restriction, or axiom. On the other hand, semantic changes include changes such as generalization or specialization of a concept or a property, increase or decrease of descriptiveness of a concept or a property, changes to restrictiveness of a property, etc. The semantic changes are the result of one or more structural changes. For instance, adding or removing a sub-class relation could move a concept up or down in the concept hierarchy making it more generalized or specialized. In this paper, we mainly analyze structural changes, which could lead to semantic changes.

Adding a new class or a property to an ontology generally does not cause problems in the data because there is no data loss. However, as it has been seen in a recent analysis of DBpedia [13], new concepts and properties can introduce duplicates in the ontology. These duplicates (i.e., concepts or properties that have the same meaning but different identifiers) decrease the conciseness of the ontology and further make it harder to query or understand data. Adding a sub-class or sub-property can introduce some undesired effects if not done in a careful manner. When a new sub-property relationship is defined, one should take care that the domain and range restrictions of the sub-property are compliant with the ones of the super-property. Otherwise, undesired facts can be inferred by a reasoner. Further, if a data provider or an application used the given property without having this new hierarchical relation in mind, they can get undesired results.

3   Approach

This sections describes the approach followed in order to carry out the proposed analysis. For doing this, the section explains first the ontology selection process (Sect. 3.1) and then the data extraction implementation details (Sect. 3.2).

3.1   Ontology Selection

For the selection of the ontologies several criteria were considered: (i) wide usage in the datasets in the LOD Cloud, (ii) availability of multiple versions of the ontology, (iii) whether the ontology was developed collaboratively, and (iv) the process used for collaborative development.

Ontology	Versions		Timespan
	Count	Range	
DBpedia	12	3.2 ~ 2016-04	2008/10 ~ 2016/10
Schema.org	24	0.91 ~ 2.2	2012/04 ~ 2015/11
W3C PROV-O	7	Initial – W3C Rec	2012/05 ~ 2015/01
FOAF	10	Initial – 0.99	2005/04 ~ 2014/01

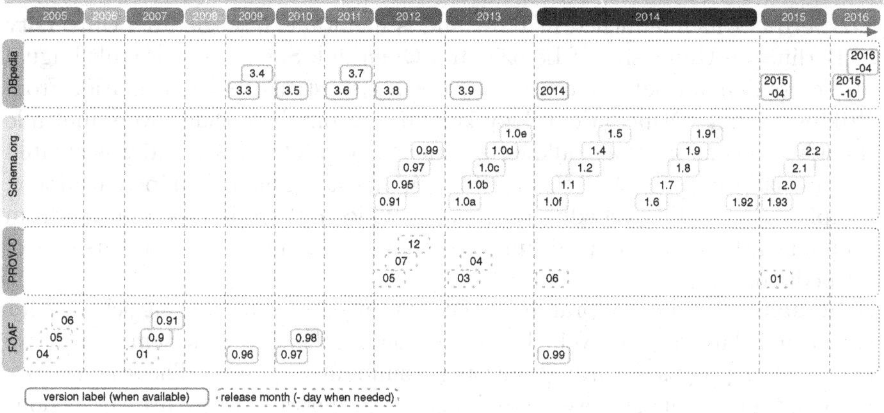

Fig. 1. Selected ontologies

For checking the usage of the ontologies, we analyzed the data from the LOD Cloud State[1] which describes the usage of terms from a given ontology in the LOD Cloud. We used the Linked Open Vocabularies (LOV[2]) [14] dataset to verify the availability of multiple versions of a given vocabulary. We only

---

[1] http://lod-cloud.net/state/#terms.
[2] http://lov.okfn.org.

considered ontologies with at least 5 versions and spanning at least a 3 year duration. Then we checked if the ontology was developed in a collaborative way with a large group of people. Finally, we tried to include ontologies developed following different processes of different organizations such as W3C, DBpedia Community, etc. Figure 1 shows a summary of the 4 ontologies selected (53 versions).

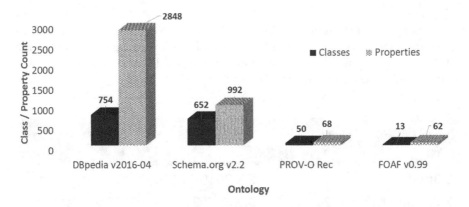

Fig. 2. Class and property counts of selected ontologies.

- **DBpedia**[3] which launched in 2007, and is maintained by the Free University of Berlin, the University of Leipzig and OpenLink Software, is a multilingual and cross-domain dataset created by extracting structured information from Wikipedia. The DBpedia ontology is a general ontology that covers multiple domains. It consists of a shallow class hierarchy of 320 classes and was created by manually deriving 170 classes from the most common Wikipedia infobox templates in the English edition. It also includes 750 properties resulting from mapping attributes from within these templates. There are 11 versions of DBpedia ontology.
- **Schema.org**[4] is a collaborative initiative that aims at promoting schemes for structured data on the Web. It consists on a collection of schemes that are extended or proposed through an open community process. The schemes are a set of 'types' that have associated a set of properties. There are 26 versions of schema.org registered in LOV.
- **PROV-O**[5] is a W3C recommendation to represent and interchange provenance information generated in different systems and under different contexts. It can also be specialized to cover particular applications and domains. This ontology implements the PROV Data Model[6]. There are 8 versions of PROV-O registered in LOV.

---

[3] http://dbpedia.org/ontology/.

[4] http://schema.org/.

[5] http://www.w3.org/ns/prov.

[6] https://www.w3.org/TR/prov-dm/.

- **FOAF**[7], which stands for "Friend of a Friend", is devoted to describe people and their relations on the Web. The core part describes characteristics of people and social groups and the social web part includes information about web accounts, address books and other web-based activities. FOAF is one of the most reused ontologies. There are 10 versions of FOAF registered in LOV.

Figure 2 shows the number of classes and properties of the latest of version of the four ontologies that were studied in this paper. As it can be seen from the chart, the four ontologies vary a lot in size. The DBpedia ontology is the largest when it comes to the number of classes and properties while Schema.org has a comparable number of classes. Nevertheless, Schema.org has considerably smaller number of properties compared to DBpedia. For example, the class to property ratio in DBpedia is 1:3.78 while in Schema.org it is 1:1.52. PROV-O and FOAF are relatively small in size compared to the other two.

3.2 Data Extraction Process

This section describes how this analysis was implemented. For discovering and extracting the versions of the different ontologies, we used LOV. LOV stores information about more than 500 vocabularies and several versions for each of them. We used the SPARQL querying facility in LOV to discover the versions of a given vocabulary and also to extract them.

For the analysis itself, we used, the Loupe [15] tool, a java-based online tool for inspecting RDF datasets that extracts statistics and data patterns in a given dataset. Loupe allows to easily analyze datasets by creating virtual SPARQL endpoints via a Dockerized Virtuoso instance and a set of parameterized SPARQL queries. In this paper, we used Loupe for extracting all the information about classes, properties, and also subclass and subproperty relations in each of the vocabulary versions and also for comparing subsequent versions to identify classes and properties that are added or removed.

4 Analysis

This section presents the results obtained from the study carried out over the four ontologies. In order to provide more context about the collaborative aspect of the development of each ontology, Sect. 4.1 discusses the ontology development process, Sect. 4.1 provides some notions about how the community was involved in the development, Sect. 4.3 analyses the changes of each ontology in detail and, finally, Sect. 4.4 compares the overall trends of the evolution of both classes and properties of the four ontologies studied.

---

[7] http://xmlns.com/foaf/0.1/.

4.1   Ontology Development Process

- The **DBpedia** community uses a wiki-based approach for developing the ontology. The DBpedia mappings wiki[8] provides guidance and templates for editing the DBpedia ontology mainly focusing on how to add classes and properties. Any community member with sufficient privileges can make modifications to the ontology. DBpedia provides some tool support for exploring the ontology and validating it after a modification. Until recently (February, 2015), it seems that DBpedia ontology issues have been tracked in an ad-hoc manner. But currently DBpedia uses an issue tracker[9] to raise and follow the progress of the DBpedia ontology issues.
- **Schema.org** allows two types of extensions to be made by the community collaboratively, namely 'hosted' and 'external' extensions. In both cases typically subclasses and properties are added to the core schema. While hosted extensions are managed and reviewed as part of the Schema.org project, the external ones are managed and reviewed by other groups.
- **PROV-O** was developed in accordance to the process[10] defined in the W3C for developing W3C recommendations. The W3C working group discussed the details of the ontology during the weekly teleconferences and also during the face-to-face meetings. Technical decisions are solved using consensus or a voting-based process. Any member can raise issues which are recorded in the issue tracker and discussed in the group to come up with resolutions.
- **FOAF** has been built by a collaborative effort of the users registered in the mailing list.[11] The vocabulary intends to be pragmatic and simple and to allow particular extensions. FOAF considers the stability of individual vocabulary terms, instead of the specification as a whole. Terms progress through the categories 'unstable', 'testing' and 'stable'. Older terms might be considered'archaic' which also allows them to become modern again.

4.2   Community

- The **DBpedia** community is the largest of the four communities that developed the selected ontologies in this study. It is also important to note that the DBpedia community is clustered into several sub-communities, *i.e.*, DBpedia language chapters, and has members from more than over 20 countries. In 2014, the DBpedia Association was founded to support DBpedia and the DBpedia community. As of November 2016, there are 488 members of the DBpedia community with write access to the ontology wiki so that they can introduce changes to the ontology. Out of those members, 14 have been active within the month of November 2016. The majority of the active members of the DBpedia ontology editors come from academia.

---

[8] http://mappings.dbpedia.org/index.php/How_to_edit_the_DBpedia_Ontology.
[9] https://github.com/dbpedia/ontology-tracker/.
[10] https://www.w3.org/2015/Process-20150901/.
[11] foaf-dev@lists.foaf-project.org.

- The **Schema.org** community is reasonably large with 48 contributors with commits to the Git repository where the ontology is maintained as of November 2016. In addition, many have indirectly contributed through the mailing list discussions, issue tracker, etc. Most of them are industrial practitioners coming from the sponsoring companies, *i.e.*, Google, Microsoft, Yahoo and Yandex. In April 2015, the Schema.org community has formed the W3C Schema.org Community Group. Schema.org also has a steering group that consists of representatives of the sponsor companies, a representative of the W3C and a small number of individuals who have contributed substantially to Schema.org.
- The **PROV-O** ontology was developed by the W3C Provenance Working Group which consists of 59 working group members. The working group had a mix of both academic and industrial participants. The "PROV-O: The PROV Ontology" W3C Recommendation lists three editors and 7 contributors who have directly contributed to developing the ontology.
- The **FOAF** ontology is developed by Dan Brickley and Libby Miller with the contributions from the members of the FOAF mailing list and the W3C Semantic Web Interest Group.

4.3   Ontology Change Analysis

In the following, an analysis of the results obtained for each of the observed ontologies is provided. For each ontology, we provide a summary of changes in a table which illustrates the number ($\#$) of classes and properties in each version of the ontology along with structural changes such as addition ($+$) or removal ($-$) of classes and properties compared to the previous version. Further, we also provide what is the effective change of both additions and removals, i.e., the difference (Δ) compared to the previous version.

More information about the ontology changes such as which classes and properties are added in each version of the ontology is available in an external wiki page[12].

- The **DBpedia** ontology changes from version 3.2 (2008) until version 2016-04 (2016) are shown in Table 1. The ontology is gradually growing with respect to the number of classes and properties. However, when we look in detail, for instance, how many terms are added and removed, it can be seen that a large number of classes and properties are removed as well.
  When we analyze the classes removed, we could find classes that do not follow the proper naming convention such as `dbo:bibo:Book`, or `dbo:Bullfighter` or duplicates of existing classes such as `dbo:Pornstar` (which is a duplicate of `dbo:AdultActor` that already exists). The same happens with the properties that are removed. In version 3.5, we can see that 1,198 properties (approximately half) are removed. This is because of a change in the convention of URI generation. For instance, before version 3.5, there were a lot of properties

---

[12] https://github.com/nandana/loupe/wiki/Ontology-Changes.

Table 1. DBpedia - evolution of classes and properties

Version	OWL Class				RDF Property				Object Prop.			Datatype Prop.		
	#	Δ	(-)	(+)	#	Δ	(-)	(+)	#	(-)	(+)	#	(-)	(+)
3.2/3	174				720				384			336		
3.4	204	30	-2	32	2168	1448	-271	1719	1144	-139	899	1024	-132	820
3.5	255	51	-6	57	1274	-894	-1198	304	601	-673	130	673	-525	174
3.6	272	17	0	17	1335	61	-37	98	629	-26	54	706	-11	44
3.7	319	47	-1	48	1643	308	-17	325	750	-6	127	893	-11	198
3.8	359	40	-1	41	1775	132	-3	135	800	-1	51	975	-2	84
3.9	529	170	-1	171	2333	558	-8	566	927	-6	133	1406	-2	433
2014	683	154	-5	159	2795	462	-46	508	1079	-9	161	1716	-37	347
2015-04	735	52	-5	57	2819	24	-103	127	1098	-23	42	1721	-80	85
2015-10	739	4	-5	9	2833	14	-9	23	1099	-3	4	1734	-6	19
2016-04	754	15	0	15	2849	16	-2	18	1103	-1	5	1746	-1	13

Table 2. DBpedia - Triples of removed properties

Property	Last Version	Triples in DBpedia ES 2016-04	Triples in DBpedia IT 2016-04
dbo:buriedPlace	2014	4519	0
dbo:diseasesdb	2014	4346	0
dbo:emedicineTopic	2014	1977	0
dbo:foundingPerson	2015-04	2158	0
dbo:medlineplus	2014	3300	0
dbo:coordinates	2015-10	0	180
dbo:score	2015-10	0	26873

in the form (http://dbpedia.org/ontology/Athlete/formerTeam) which were changed to (http://dbpedia.org/ontology/formerTeam). However, the reasons for removal or any other provenance metadata are not documented in any place. Furthermore, we can see that from the 754 classes added to the ontology 514 classes (72%) are only used by less than 5 datasets[13]. With respect to data quality, we analyzed if the instances of removed classes or triples containing some of the removed properties were introduced to data after their removal. For this purpose, we analyzed the LOD Cache[14] dataset and the different language-specific datasets of DBpedia. We could find examples for both cases as illustrated in Tables 2 and 3. Further, we noticed several classes and properties that have been added and removed several times within the period we analyzed. Since these classes and properties are unstable, it makes it difficult to the users of the ontology to decide whether or not to use them to annotate their data. Unlike FOAF or Schema.org, the DBpedia ontology

---

[13] http://nandana.github.io/dbpedia/2015-10/class-langs.html.
[14] https://datahub.io/dataset/openlink-lod-cache.

does not annotate the status or the maturity of its terms. Few examples of unstable classes and properties (*i.e.*, the ones that are removed and introduced several times) are illustrated in Fig. 3. We could also find that when classes and properties were added, some duplicates were also introduced, *i.e*, classes referring to the same concept and also properties referring to the same relationship. Some examples are classes such as dbo:AdultActor and dbo:PornStar classes and properties such as dbo:color and dbo:colour or dbo:foundingDate and dbo:formationDate.

There are several key issues in the DBpedia ontology. First, being quite a large ontology, it is not sufficiently modular; its large monolithic nature hinders the proper reuse of it. As a consequence, the ontology editors introduce duplicate classes and properties for the same concepts and relations because they are not aware of or could not find the appropriate existing term. Another factor that contributes to the same problem is the minimum documentation in some classes or properties. Out of 2,849 properties, only 556 properties have a comment or description associated with them. Most of these problems exist because of the rather relaxed editorial process in minimal review and governance.

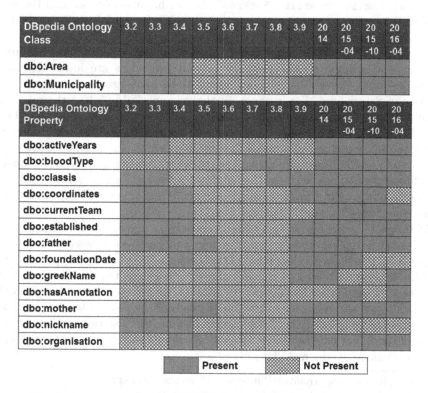

Fig. 3. An excerpt of unstable DBpdia ontology classes and properties

Table 3. Instances of removed classes

Class	Latest Version	instances in LOD Cache	instances in DBpedia IT 2016-04
dbo:Bullfighter	2015-04	2	-
dbo:Comics	2014	256	2241
dbo:Imdb	2015-04	3	-
dbo:Installment	2015-04	601	-
dbo:Pornstar	3.9	2	-

- **Schema.org** is also a fairly large vocabulary that releases versions of the ontology frequently. Table 4 shows the ontology changes of 25 different versions of Schema.org since 2012. Similar to the DBpedia ontology, the Schema.org ontology is also gradually growing. However, in contrast to the DBpedia ontology we can notice a lower number of removals of concepts or properties. The only class that was removed `schema:OnSitePickup` was later reintroduced into the ontology. The four properties that have been removed `schema:oponent`[15], `schema:supercedes`[16], `schema:numberofEmployees`[17], and `schema:isAcccessibleForFree`[18] are all because of typos and the reasons for removing them are documented either in the issue tracker or the mailing list.

  However, it is not the same when we analyze the subclass and subproperty relationships. As the new classes and properties are introduced and the hierarchy grows, the relations are changed frequently in the recent versions of Schema.org. For instance, in the version "2015-05-12", the relationship `schema:BookSeries` is a subclass of `schema:Series` is removed. This is because a new hierarchy is introduced as `schema:CreativeWorkSeries` which is a subclass of `schema:CreativeWork` and `schema:BookSeries` is moved to be a subclass of `schema:CreativeWorkSeries`. This type of changes could have implications on the applications which are not aware of the change and expect the instances of `schema:BookSeries` to be a subclass of `schema:Series`.

- **PROV-O** is a fairly small-sized ontology compared to DBpedia or Schema.org. It has a very focused scope and has been developed following the W3C process with close interactions between the working group members. Table 5 shows the ontology changes of seven versions of the PROV-O ontology from 2012 until 2015. In the case of PROV-O, we can see a lot of changes both as additions and removals in the initial phase of the process but less changes in the later phases of the development. In fact, the last two versions of the ontology do not have additions or removals of concepts or properties and only improvements to the metadata.

---

[15] https://lists.w3.org/Archives/Public/public-vocabs/2014Apr/0289.html.
[16] https://github.com/schemaorg/schemaorg/issues/101.
[17] https://github.com/schemaorg/schemaorg/issues/252.
[18] https://github.com/schemaorg/schemaorg/issues/508.

Table 4. Schema.org - evolution of classes and properties

Version		RDFS Class				RDF Property				Subclass			Subprop.		
Date	#	V.	Δ	(-)	(+)	#	Δ	(-)	(+)	#	(-)	(+)	#	(-)	(+)
2012-04-27	0.91	302				286				317			0		
2012-06-26	0.95	391	89	0	89	465	179	0	179	413	0	96	0	0	0
2012-07-26	0.97	393	2	0	2	466	1	0	1	415	0	2	0	0	0
2012-11-08	0.99	416	23	0	23	544	78	0	78	438	0	23	0	0	0
2013-04-05	1.0a	428	12	0	12	581	37	0	37	451	0	13	0	0	0
2013-07-24	1.0b	428	0	0	0	582	1	0	1	451	0	0	0	0	0
2013-08-07	1.0c	531	103	0	103	627	45	0	45	554	0	103	0	0	0
2013-11-19	1.0d	552	21	0	21	675	48	0	48	577	-1	24	0	0	0
2013-12-04	1.0e	558	6	0	6	711	36	0	36	583	0	6	0	0	0
2014-02-05	1.0f	558	0	0	0	711	0	0	0	583	0	0	0	0	0
2014-04-04	1.1	582	24	-1	25	777	66	0	66	607	1	25	1	0	1
2014-04-16	1.2	585	3	0	3	792	15	0	15	610	0	3	1	0	0
2014-05-16	1.4	585	0	0	0	794	2	-1	3	627	0	17	1	0	0
2014-05-27	1.5	585	0	0	0	798	4	0	4	627	0	0	1	0	0
2014-06-16	1.6	588	3	0	3	803	5	0	5	632	-2	7	36	-1	36
2014-07-08	1.7	589	1	0	1	806	3	0	3	633	0	1	36	0	0
2014-07-28	1.8	590	1	0	1	806	0	0	0	634	0	1	36	0	0
2014-08-19	1.9	593	3	0	3	816	10	0	10	636	-1	3	42	0	6
2014-09-12	1.91	593	0	0	0	816	0	-1	1	637	0	1	44	0	2
2014-12-11	1.92	618	25	0	25	878	62	0	62	663	-3	29	55	0	11
2015-02-04	1.93	620	2	0	2	891	13	-1	14	665	0	2	55	0	0
2015-05-12	2.0	638	18	0	18	965	74	0	74	676	-17	28	62	0	7
2015-08-06	2.1	645	7	0	7	976	11	-1	12	683	-2	9	63	0	1
2015-11-05	2.2	652	7	0	7	992	16	0	16	682	-10	9	69	0	6

When analyzing the classes added and removed in different versions, is seems they were done depending on the scope of the ontology agreed by the group. For example, there are several classes which are removed in the version "2012-07-24" such as prov:Dictionary or prov:Insertion are re-added in the version "2013-04-30". An expansion of scope can be seen in the version "2012-07-24" where one can find several concepts related to influence such as prov:Influence, prov:AgentInfluence, prov:ActivityInfluence, and prov:EntityInfluence.

- **FOAF** is the smallest and the oldest among the ontologies that have been analyzed. Table 6 shows the ontology changes of ten versions of the FOAF ontology since 2005 until 2014. Notably the FOAF vocabulary has not removed any classes or properties from the previous versions. Though it seems that two object properties were removed, those refer to object properties which were transformed into datatype or annotation properties (namely foaf:membershipClass in v2007-01-14 and foaf:myersBriggs in v2009-12-15v2005-06-03) (Fig. 6).

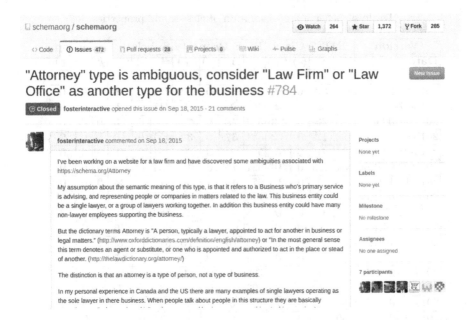

Fig. 4. Issue tracking in Schema.org

Class	v0.97	v0.98	v0.99
foaf:Agent			
foaf:Person		Stable	
foaf:Group			
foaf:Organization			
foaf:Document			
foaf:Image			
foaf:PersonalProfileDocument		Testing	
foaf:OnlineAccount			
foaf:Project			
foaf:OnlineChatAccount			
foaf:OnlineEcommerceAccount		Unstable	
foaf:OnlineGamingAccount			
foaf:LabelProperty	N/A		

Fig. 5. Evolution of FOAF classes

With respect to subclass relationships, there are some relations that have been removed. These are mainly subclass relations to external ontologies. For instance, in "2007-01-14" and "2009-12-15" the subclass relations to classes from "http://xmlns.com/wordnet/1.6/" were removed. In version "2014-01-14", the subclass relation to pim:Person was removed.

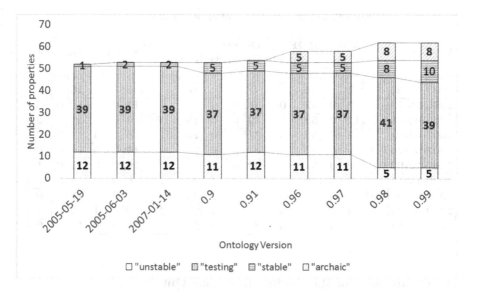

Fig. 6. Evolution of FOAF properties

FOAF uses the `term_status` property from the SemWeb Vocab Status ontology[19] to annotate the level of stability of a class or a property into four categories, i.e., *Unstable, Testing, Stable, Archaic*. As the FOAF ontology evolved through different versions, it has moved the status of the vocabulary terms from *Unstable* to *Testing*, and from *Testing* to *Stable*. Further, when a vocabulary term becomes obsolete, the FOAF ontology has labeled them as *Archaic*. Figure 5 shows how the status of the classes of the FOAF ontology has evolved during its last three versions and Fig. 6 illustrates the evolution of the status of the FOAF properties during its last nine versions.

Table 5. PROV-O - evolution of classes and properties

Version	OWL Class				RDF Property				Object Prop.			Datatype Prop.		
	#	Δ	(-)	(+)	#	Δ	(-)	(+)	#	(-)	(+)	#	(-)	(+)
2012-05-03	38				60				55			5		
2012-07-24	30	-8	-16	8	52	-8	-22	14	46	-20	11	6	-2	3
2012-12-11	30	0	-1	1	50	-2	-3	1	44	-3	1	6	0	0
2013-03-12	30	0	0	0	50	0	0	0	44	0	0	6	0	0
2013-04-30	50	20	0	20	68	18	0	18	59	0	15	9	0	3
2014-06-07	50	0	0	0	68	0	0	0	59	0	0	9	0	0
2015-01-11	50	0	0	0	68	0	0	0	59	0	0	9	0	0

---

[19] https://www.w3.org/2003/06/sw-vocab-status/.

Table 6. FOAF - evolution of classes and properties

Version		OWL Class #	(+)	RDF Property #	(+)	Object Property #	(-)	(+)	Datatype Property #	(+)	Subclass #	(-)	(+)	Subprop. #	(-)	(+)
Date																
2005-04-03		12		52		0			0		15			10		
2005-05-19		12	0	52	0	0	0	0	0	0	15	0	0	10	0	0
2005-06-03		12	0	53	1	32	0	32	19	19	15	0	0	11	-1	2
2007-01-14		12	0	53	0	31	-1	0	20	1	16	-3	4	11	0	0
2007-05-24	0.9	12	0	53	0	31	0	0	20	0	16	0	0	11	0	0
2007-10-02	0.91	12	0	54	1	32	0	1	20	0	16	0	0	12	0	1
2009-12-15	0.96	12	0	58	4	32	-1	1	24	4	9	-7	0	12	0	0
2010-01-01	0.97	12	0	58	0	32	0	0	24	0	9	0	0	12	0	0
2010-08-09	0.98	13	1	62	4	33	0	1	27	3	10	0	1	13	0	1
2014-01-14	0.99	13	0	62	0	33	0	0	27	0	9	-1	0	13	0	0

4.4   Comparison of the Evolution of Four Ontologies

When comparing the evolution of the ontology classes (see Fig. 7), it is clear that
the number of classes has grown gradually in all the four ontologies and removals
have been minimal. During 2013–2014, there is a steep growth in the DBpedia
ontology while similar activity can be seen in Schema.org around its 1.0c release.

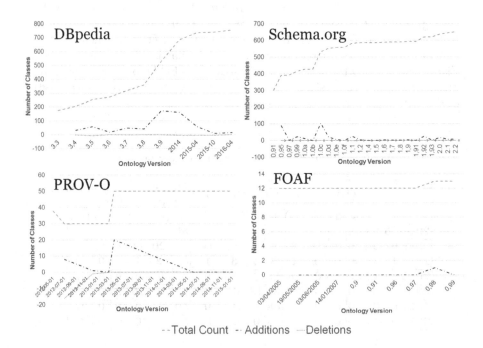

Fig. 7. Comparison of evolution of classes

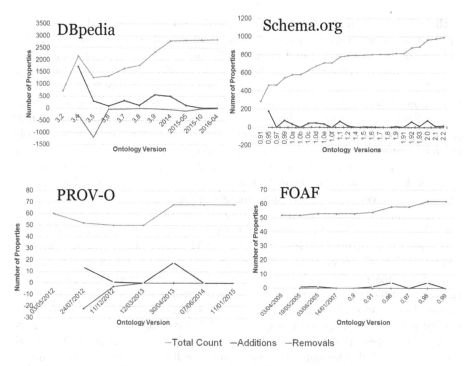

Fig. 8. Comparison of evolution of properties

PROV-O has one big change before becoming a Candidate Recommendation and FOAF, being the smallest of the four, has not suffered from big changes to its classes.

When comparing the evolution of the ontology properties (see Fig. 8), it can be seen that there are a lot of changes in properties both in additions as well as in removals compared to the changes in classes. Specially in DBpedia we can see a large number of property removals. Nevertheless, the more recent versions seem to have lesser changes in both additions and removals. In Schema.org, the number of properties has gradually grown through versions with a minimal number of removals. Similar to the changes in classes, in PROV-O there are several changes before it becomes a Candidate Recommendation and then it becomes stable. There are a small number of additions in the FOAF properties in the last few versions.

5   Conclusions and Future Work

Ontology changes are inevitable; the knowledge of a domain of interest changes and ontology evolution is needed for making sure that those changes are adapted by the ontology. We have observed that these changes depend also on the characteristics of the ontology development process and on the community involved. For instance, the DBpedia ontology follows a bottom-up approach by trying to fit the ontology to the infoboxes from Wikipedia; on the other hand, Schema.org and PROV follow a top-down approach by expanding its scope to different areas while studying the domain and including the most relevant concepts and relations. It's also important to note that the level of manual curation involved in each process varies a lot and it's evident that a single ontology evolution methodology does not fit all different processes and communities. The analysis shows that in communities such as Schema.org, where industrial participation is dominant and there is a strong commercial interest, the community tends to follow more rigorous editorial processes and governance procedures. Furthermore, it shows that the tools used by such communities, for instance, version control systems, issue trackers, implementation reviews, etc., are largely influenced by the software development tools and processes (Fig. 4).

Furthermore, during the study we noted that it is hard to find practical guidelines and best practices for ontology evolution. Even though some guidelines such as "Principles of Good Practice for Managing RDF Vocabularies and OWL Ontologies" [16] exist, they do not provide practical guidance for managing ontology changes in different scenarios (e.g., top-down vs bottom-up) or recipes on how to implement those guidelines in such scenarios. If more guidelines were provided on how to manage ontology changes, that would help to reduce the quality issues introduced by the changes while evolving an ontology. Furthermore, it is evident from the study that even though there are several theoretical frameworks available for ontology evolution in the literature, none of the studied ontologies seems to follow such frameworks.

From the analysis carried out we can draw as a suggestion that there is a necessity of incorporation of a rigorous editorial process and governance to DBpedia as well as a set of tools to facilitate the proper evolution of the ontology. As DBpedia makes use of crowdsourcing at many different stages and includes people from different levels and areas of knowledge about the ontology, having a proper editorial process becomes very important.

During the analysis, we have noticed that ontology evolution becomes more challenging when the ontology becomes larger and the development is performed in a collaborative manner. This is apparent in the DBpedia ontology; in this case, modularizing the ontology would help, so that it is easier to understand and extend it. For instance, in the DBpedia ontology there are more than 2,800 properties and a DBpedia member adding a new property needs to know all of them to avoid duplicates. However, if those were in a more modular manner it would help to avoid these problems.

In the same line, it would be helpful to have tools that could help in the process of ontology evolution. At the moment, there are tools that verify the consistency of an ontology after a change [17]. However, other functionalities and techniques such as autocompletion ones, which look for similar existing terms in the ontology both using lexical or semantic similarity, can be used to avoid duplicate terms.

Furthermore, in evolving ontologies it is important to have provenance information about how the terms have evolved. Thus, in addition to the generic metadata, such as the label or description, it would be useful to add information such as when the term was added, who was the editor who added it, why a term was deleted, etc. and references to any discussions about the reasons for the change.

As future work, we plan to extend our analysis to more change operations in ontologies including modifications to axioms, such as restrictions, and their impact on data quality. We also plan to produce a set of best practices for minimizing the data quality issues caused by ontology evolution based on the lessons learned.

Furthermore, we plan to do an extended analysis of the impact of the evolution of the ontologies discussed in this study using the datasets available in the Linked Open Data cloud that use such ontologies.

Acknowledgments. This work was funded by the BES-2014-068449 grant under the 4V project (TIN2013-46238-C4-2-R).

References

1. Gruber, T.R., et al.: A translation approach to portable ontology specifications. Knowl. Acquisition **5**(2), 199–220 (1993)
2. Noy, F.N., Klein, M.: Ontology evolution: not the same as schema evolution. Knowl. Inf. Syst. **6**(4), 428–440 (2004)
3. Stojanovic, L.: Methods and tools for ontology evolution. Master's thesis, Karlsruhe Institute of Technology, Karlsruhe, Germany (2004)
4. Stojanovic, L., Maedche, A., Motik, B., Stojanovic, N.: User-driven ontology evolution management. In: Gómez-Pérez, A., Benjamins, V.R. (eds.) EKAW 2002. LNCS (LNAI), vol. 2473, pp. 285–300. Springer, Heidelberg (2002). doi:10.1007/3-540-45810-7_27
5. Palma, R.: Ontology metadata management in distributed environments (2009)
6. Zablith, F.: Evolva: a comprehensive approach to ontology evolution. In: Aroyo, L., Traverso, P., Ciravegna, F., Cimiano, P., Heath, T., Hyvönen, E., Mizoguchi, R., Oren, E., Sabou, M., Simperl, E. (eds.) ESWC 2009. LNCS, vol. 5554, pp. 944–948. Springer, Heidelberg (2009). doi:10.1007/978-3-642-02121-3_87
7. Haase, P., Harmelen, F., Huang, Z., Stuckenschmidt, H., Sure, Y.: A framework for handling inconsistency in changing ontologies. In: Gil, Y., Motta, E., Benjamins, V.R., Musen, M.A. (eds.) ISWC 2005. LNCS, vol. 3729, pp. 353–367. Springer, Heidelberg (2005). doi:10.1007/11574620_27
8. Pinto, H.S., Staab, S., Tempich, C.: DILIGENT: towards a fine-grained methodology for DIstributed, Loosely-controlled and evolvInG Engineering of oNTologies. In: Proceedings of the 16th Eureopean Conference on Artificial Intelligence, ECAI 2004, including Prestigious Applicants of Intelligent Systems, PAIS 2004, Valencia, Spain, 22–27 August 2004, pp. 393–397 (2004)
9. Dragoni, M., Ghidini, C.: Evaluating the impact of ontology evolution patterns on the effectiveness of resources retrieval. In: Proceedings of the 2nd Joint Workshop on Knowledge Evolution and Ontology Dynamics (EvoDyn-2012), vol. 890 (2012)
10. Abgaz, Y.M., Javed, M., Pahl, C.: Analyzing impacts of change operations in evolving ontologies. In: Proceedings of the 2nd Joint Workshop on Knowledge Evolution and Ontology Dynamics (EvoDyn-2012), vol. 890 (2012). CEUR-WS.org

11. Noy, N.F., Musen, M.A., et al.: PROMPTDIFF: a fixed-point algorithm for comparing ontology versions. In: Proceedings of the Eighteenth National Conference on Artificial Intelligence vol. 2002, pp. 744–750 (2002)

12. Qin, L., Atluri, V.: Evaluating the validity of data instances against ontology evolution over the semantic web. Inform. Softw. Technol. **51**(1), 83–97 (2009)

13. Mihindukulasooriya, N., Rico, M., García-Castro, R., Gómez-Pérez, A.: An analysis of the quality issues of the properties available in the Spanish DBpedia. In: Puerta, J.M., Gámez, J.A., Dorronsoro, B., Barrenechea, E., Troncoso, A., Baruque, B., Galar, M. (eds.) CAEPIA 2015. LNCS (LNAI), vol. 9422, pp. 198–209. Springer, Cham (2015). doi:10.1007/978-3-319-24598-0_18

14. Vandenbussche, P.Y., Atemezing, G.A., Poveda-Villalón, M., Vatant, B.: Linked Open Vocabularies (LOV): a gateway to reusable semantic vocabularies on the web. Semantic Web J. **8**(3), 437–452 (2016)

15. Mihindukulasooriya, N., Poveda-Villalón, M., García-Castro, R., Gómez-Pérez, A.: Loupe-an online tool for inspecting datasets in the linked data cloud. In: Demo at the 14th International Semantic Web Conference, Bethlehem, USA (2015)

16. Kendall, E., Novacek, V.: Principles of Good Practice for Managing RDF Vocabularies and OWL Ontologies, March 2008

17. Haase, P., Stojanovic, L.: Consistent evolution of OWL ontologies. In: Gómez-Pérez, A., Euzenat, J. (eds.) ESWC 2005. LNCS, vol. 3532, pp. 182–197. Springer, Heidelberg (2005). doi:10.1007/11431053_13

Towards Ontology-Based Event Processing

Riccardo Tommasini[1,2], Pieter Bonte[1,2(✉)], Emanuele Della Valle[1,2],
Erik Mannens[1,2], Filip De Turck[1], and Femke Ongenae[1,2]

[1] imec, Ghent University, Ghent, Belgium
{pieter.bonte,erik.mannens,filip.deturck,femke.ongenae}@ugent.be
[2] DEIB, Politecnico di Milano, Milan, Italy
{riccardo.tommasini,emanuele.dellavalle}@polimi.it

Abstract. The rapid change and heterogeneity of today's generated data calls for real-time decision making systems that can cope with the presented heterogeneity. In this paper, we present an Ontology Based Event Processing system that bridges the gap between ontology-based reasoning and event processing. We propose both a language and an architecture to perform event processing over abstract ontology concepts. This allows to perform efficient temporal reasoning, while the high-level ontological definitions reduce the need for knowledge of the underlying data structure in complex domains.

Keywords: Stream Processing · Semantic Web · Stream Reasoning · Complex Event Processing

1 Introduction

In domains like Social Media, Financial Markets and Internet of Things (IoT), information is traditionally represented as data streams, i.e. unbounded sequences of data, or events, i.e. notifications about happened facts. Stream Reasoning (SR) [5] investigates how Semantic Web and Stream Processing technologies can be combined to make decision making systems work in real-time, across multiple data sources. SR investigates how to exploit the time ordering of data streams to perform deductive and temporal reasoning on the fly.

In order to clarify this domain, consider the following example: We are interested to identify the presence of fire in a room, but there is no way to detect it directly. Instead, the room contains sensors to detect the presence of smoke and measure the temperature. In this case, a data stream is a timestamped sequence of numbers representing the average temperature in the room; while an event is a notification about the detection of smoke. The data and events arise from different types of sensors. This heterogeneity impedes to perform queries across these data sources. Another obstacle comes from the domain complexity. In presence of fire the temperature will be higher, but how can we distinguish abnormal temperatures from normal ones? And, what if we had different rooms? This kind of information represent background knowledge that our decision making system has to combine with live data, in order to obtain an answer. Finally, assuming

© Springer International Publishing AG 2017
M. Dragoni et al. (Eds.): OWLED-ORE 2016, LNCS 10161, pp. 115–127, 2017.
DOI: 10.1007/978-3-319-54627-8_9

that we finally detect both smoke and abnormal temperature events, we have to relate them in time.

The presented example calls for an approach that solves data variety, that combines data with background knowledge, that deducts related information and operates temporal reasoning combining data streams from sensors and events. We name this approach Ontology-Based Event Processing (OBEP). For the best of our knowledge, there is no approach in the SR state of the art that tries to do so. Temporal extensions of deductive reasoning extends the ontological language with time relations and, thus, easily diverges into intractability. Semantic Complex Event Processing is limited to a semantic description of events and does not focus on the processing.

In this paper, we propose an approach for OBEP that operates the event processing a-posteriori above high level concepts deduced through deductive reasoning, but without including time relations at ontological level. The contribution of this work are: (i) a requirement analysis for an OBEP system to satisfy; (ii) a syntax named DELP, i.e. Description Logic Event Processing, to express information needs as the one presented in the example; (iii) an architecture that bridges the gap between event processing to capture temporal relations and event descriptions based on Semantic Web technologies and; (iv) a prototype that proves the feasibility of the approach.

The rest of the paper is structured as follows: Sect. 2 describes the related works. Section 3 describes the use case that is used throughout the paper. Section 4 introduces the Description Logic Event Processing (DELP) language we constructed, while Sect. 5 describes our OBEP system that implements the abstracted event processing. Section 6 concludes the paper and elaborates on the future research directions.

2   Background and Related Work

In this section we present the background knowledge required to understand the content of the paper and the relevant related work.

Stream Processing engines are systems capable to process potentially infinite sequences of data. Two main approaches exits to this extent:

- Data Stream Management Systems (DSMS) extend Data Base Management Systems by introducing stream-to-relation operators, e.g. Windows, that allow the transition between streaming and static data. Queries are continuously evaluated over finite portions of the data streams selected by the means of these operators.
- Complex Event Processing (CEP) engines exploit time-aware operators to detect patterns over infinite sequences of incoming events. The user specifies reaction rules that are concerned with the invocation of actions in response to events and actionable situations. These rules specify a pattern over the incoming data, e.g. A followed-by B, by using a declarative query language. Such a pattern is usually validated with a finite state machine. Therefore, the final complexity is at most polynomial in time and space.

Some stream processing engines offer declarative query language to operate with data streams. The event processing language (EPL)[1] is the most relevant one and it allows to (i) write window-based continuous queries to process data streams; (ii) define simple events or compositions of them (i.e. complex events) (iii) treat events as first class citizens, i.e. the operators have direct influence on the events.

Semantic technologies such as RDF, OWL and SPARQL have been used for data integration in the IoT domain [2] and Semantic Complex Event Processing (SCEP) [9].

An example of the former is MASSIF [4], i.e. an event-based semantic-enabled IoT platform consisting of multiple semantic reasoning services each fulfilling a distinct reasoning task. These services can collaborate on a high level by subscribing to the Semantic Communication Bus (SCB) and indicating the high level concepts they are interested in. The platform follows the notion of high level events, however, it does not support any temporal reasoning between these events.

An example of the SCEP is the work of Taylor et al. [9], i.e. an ontology and a system for complex event specification that, in combination with reasoning techniques, simplify the rule definitions of a target complex event processing language (e.g. EPL), eliminating the need of address manually the domain complexity. To this extent, the ontology contains language constructs and operators, e.g., *seq*, as properties and classes. This approach generalizes the query definition task enabling interoperability between different event processing engines, but it does not extend the semantics of the target query language nor does it propose a unified syntax for it.

In the SR state-of-the-art, RDF Stream Processing (RSP) engines combines semantic technologies and stream processing to perform continuous querying or complex event processing [1] over streams encoded into time-annotated RDF. EP-SPARQL [1] is the most relevant work w.r.t. ours, because it extends SPARQL 1.0 with event processing operators, i.e., *seq, equals, optionalseq, and equalsoptional*[2]. Event processing and SR is enabled over RDF Basic Graph Pattern (BGP). Complex events are defined as BGPs combined with event processing operators. As this is similar to the UNION or OPTIONAL operators in SPARQL, events are not first class citizens. Since the events are defined through BGPs, it can be devious to construct advanced event processing patterns.

Finally, temporal extension of deductive reasoning approaches such as Description Logics are worth to mention. They include time relations at ontological level, but this easily diverges into intractability and limiting the possible entailments [6].

In summary, state-of-the-art solutions in the stream processing context successfully model time relations but lack to address the data variety and the domain complexity. Semantic technologies can be used to describe these extents, but

---

[1] https://docs.oracle.com/cd/E13157_01/wlevs/docs30/epl_guide/overview.html.
[2] The semantics of these operators is similar to a left, right or full -join but their selectivity depends on how the constituents are temporally related.

existing approaches either lack to provide an unified syntax to model the full processing [9] or have limited expressiveness and do not treat events as first class citizens [1]. Finally, temporal logics are limited due to the hurdle of including time within the reasoning algorithms.

3   Use Case

In this section we introduce a simple use case that we will use in the reminder of the paper to explain our contributions.

A company wants to deploy an intelligent system to detect dangerous situations. Internally, they distinguish between three classes of conditions:

- Hazardous, i.e., situations that are dangerous for the company assets, e.g., fire or floods,
- Risky, i.e., situations that are dangerous for the complete business, e.g., information leaks or unauthorized access to restricted areas, and
- Unsafe, i.e., situations that are directly dangerous for people, e.g., fire or gas leaks.

For each dangerous situation class, different alarms are defined (e.g., sound and lights), alternative escape plans are organized and different authorities are responsible for handling the situation.

The company is interested in monitoring Unsafe situations within their buildings and the surrounding areas. To this extent, sensors for smoke detection, temperature, humidity and air quality monitoring are deployed within the building into a wireless sensor network. To monitor the surrounding areas, a public infrastructure provided by the local government is available through web APIs.

For the remainder of the paper, we will provide examples of the Unsafe situation Fire Detection. As explained in the Sect. 1, there is no direct way to sense fire, but we can assume its presence through the detection of smoke and abnormal temperature measurements within the same time interval. Many challenges arise to define such a simple rule:

(i) *Data Integration*: How can the proprietary data and those coming from external APIs be combined?
(ii) *Domain Complexity*: How can we decide if the detected temperature is abnormal?
(iii) *Temporal Relation*: How do we model the temporal relation between smoke events and abnormal temperature so we can infer the presence of fire?

4   Ontology-Based Event Processing Language

In this section, we introduce our first contribution: DELP, a syntax for Description Logic Event Processing. DELP is designed based on the definition of the following requirements elicited on the challenges presented in Sect. 3.

(R1) Semantic Event Representation [9]: this allows the integration of multiple heterogeneous sources (a) and derivation of implicit data in combination with background knowledge (b).

(R2) Event Processing [1]: this allows to combine high level ontological concepts capturing the temporal dependencies and build complex events.

(R3) First Class Citizens Events, i.e., creation and direct manipulation with language operators (e.g. pattern matching) should be possible.

(R4) Filtering and Joining: The former allows to remove irrelevant events, while the latter allows to combine events over multiple event streams to achieve intelligent decision making.

In Sect. 4.2, we show each challenge should be tackled for an OBEP system, finally in Sect. 4.3 we present the grammar of DELP and how it fulfills the requirements above.

4.1 Semantic Event Representations

In our running example, we want to derive abnormal temperature and measurements and combine them with smoke detection events. These needs are captured by challenges (i) and (ii), and call for a semantic representation of events. This need becomes clear when we analyze the domain complexity, e.g. temperature normality is different in different building areas, e.g., elevator are colder than server rooms.

Static Information Integration systems such as Ontology Based Data Access systems solve these circumstances by the means of an integrated conceptual model (ICM). The ICM enables query answering across heterogeneous data sources by the means of a common vocabulary formally specified with an ontological language, e.g. DL or OWL. The ICM of our example currently contains axioms from (1) to (5).

$$SmokeDetectionEvent \equiv \exists hasContext.(\exists observedProperty.Smoke) \qquad (1)$$

$$TemperatureEvent \equiv Observation$$
$$\sqcap (\exists observedProperty.Temperature) \qquad (2)$$

$$AbnormalTemperatureEvent \sqsubseteq TemperatureEvent \qquad (3)$$

$$ElevatorAbnormalTemperatureEvent \sqsubseteq AbnormalTemperatureEvent$$
$$\sqcap (\exists observationResult.[hasValue_{>40}])$$
$$\sqcap (\exists hasLocation.Elevator) \qquad (4)$$

$$ServerRoomTemperatureEvent \sqsubseteq AbnormalTemperatureEvent$$
$$\sqcap (\exists observationResult.[hasValue_{>20})$$
$$\sqcap (\exists hasLocation.ServerRoom) \qquad (5)$$

Data integration requires a generic data model. RDF is commonly used by the Semantic Web community to overcome the heterogeneity of static data. In our case, RDF is enough to represent the background knowledge but not to represent streams, which require RDF Streams (see Sect. 2).

Last but not least, the ICM, if combined with a reasoners, allows to exploit background knowledge to derive information that is only implicit described in the data, as the axioms (4) and (5) show.

Deciding the entailment to use for representing the ICM is a domain specific problem and a trade-off with the final system complexity. One may argue the need of a very expressive ontological language such as OWL 2 DL, that allows us to define events in a generic and concise manner and it enables to create a truly abstracted view over the events by the means of DL reasoning. Fragments like OWL RL, DL-lite, or EL++ have been shown to be interesting for Stream Reasoning use cases. At this stage, we do not discuss which restriction DELP should include. In order to express meaningful examples w.r.t. our use case we opted for OWL 2 DL[3], postponing a deep complexity study for future work.

4.2   Capturing Time Relations

In our running example, the central part represent the time relation between abnormal temperature and smoke. This need is captured by challenge (iii), that calls for event processing operators. In practice, we need to explain simple temporal pattern such as *seq*, combined with modifiers that provide enough expressiveness to capture the entire domain complexity, e.g. *not*.

Regarding time, we assume a point-based time semantics [3] for events; an event e as a pair (G, t), where G is an RDF graph containing the event statements and t is the associated timestamps. A partial ordering is established among events, i.e. events can occur at the same timestamps. Regarding the event processing, we consider the following time-aware operators:

- *seq*: (G_1, t_1) and (G_2, t_2), returns true iff the events occur and $t_1 > t_2$;
- *and*: (G_1, t_1) and (G_2, t_2), returns true when both the events occur regardless their ordering;
- *or*: (G_1, t_1) or (G_2, t_2), returns true iff at least one of the events occur;

and the following modifiers:

- *every*, forces the re-evaluation of the pattern according to its positive evaluation;
- *within*, limits the validity of the pattern by constraining its evaluation into time boundaries; and
- *not*, negates the truth value of a pattern[4].

Notably, in the state of the art, none of the existing solution implements all these operators.

---

[3] https://www.w3.org/TR/owl2-direct-semantics/.
[4] Not can be used only as a combination of other patterns.

4.3 Description Logic Event Processing

In this section, we finally explain how the event processing operators (see Sect. 4.2) are used in combination with ontological concepts.

In our example, we are interested in abnormal temperature and smoke sensor readings to detect fire. We saw in Sect. 4.2 that semantic event representation (R1) is possible in the ICM. Alternatively, high level events can be specified within a DELP query, by the means of the EVENTDECL clause (see Listing 1.4). Listing 1.1 is an example of event declaration in DELP. The Manchester syntax[5] is chosen for two reasons: it is conciser than RDF and highlights the idea of specifying events using high level abstractions. Moreover, it was combined already with SPARQL in the past [8].

```
EVENT  : OfficeAbnormalTemperaturEvent subClassOf
  AbnormalTemperaturEvent
      and (observation_result some (hasValue (hasDataValue >= 40)))
      and (hasLocation some Office))
```

Listing 1.1. Event Declaration for office abnormal temperature in DELP

Events defined through this clause are added to the TBox of ontology the reasoner uses for the inference process. Each of the defined events in DELP are translated to OWL class expressions. The translation is straight forward, since the event definition is based on the DL Manchester syntax. For example, the Office Abnormal Temperature definition in Listing 1.1 is translated to:

$$OfficeAbnormalTemperaturEvent \sqsubseteq AbnormalTemperatureEvent$$
$$\sqcap (\exists observationResult.[hasValue_{>40}])$$
$$\sqcap (\exists hasLocation.Office) \quad (6)$$

DELP exploits the time-aware operators as explained in Sect. 4.2. Listing 1.2 shows how Fire detection can be defined exploiting the temporal relation between a SmokeDetectionEvent and AbnormalTemperaturEvent.

Event processing over high level concepts (R2), an example of which is available in Listing 1.2, is enabled by the sub-clause PATTERNEXPR of the PATTERN-DECL clause. The definition of event patterns relies on user-defined ontological concepts or those already existing in an ontology.

```
NAMED EVENT  : FireEvent {
    MATCH : AbnormalTemperaturEvent SEQ : SmokeDetectionEvent WITHIN (5m)
}
```

Listing 1.2. Event Declaration for fire, based on temperature and smoke, in DELP.

---

[5] https://www.w3.org/TR/owl2-manchester-syntax/.

```
NAMED EVENT : FireEvent {
    MATCH : AbnormalTemperaturEvent SEQ : SmokeDetectionEvent^a WITHIN (5m)
    IF {
        EVENT : AbnormalTemperaturEvent { ?tmpSnsLoc : hasValue ?v}
        EVENT : SmokeDetectionEvent { ?smkSnsLoc : hasValue ?v;
            ?smokeObs ssn:observationResult ; : hasValue ?smokeLevel
            FILTER (?smokeLevel == "3"^^xsd:integer)
            }
        }
}
```

Listing 1.3. Example of event pattern with filters (R4). [a]We assume this event is already defined in the ontology.

Last but not least, the IFDECL clause enables to express filters and joins over RDF Streams. Using a SPARQL-like syntax, the user can specify a basic graph pattern to match for each event, e.g., EVENT :AbnormalTemperaturEvent in Listing 1.3, and joins that exploit a name-based notation, i.e., variables with the same name obtain the same binding (e.g., variable ?v in Listing 1.3). Filters are specified using the SPARQL 1.1 Filter clause e.g., variable ?smokeLevel in Listing 1.3.

Finally, Listing 1.4 describes a sub-portion of the DELP grammar, the full one is available at http://bit.ly/2bURXUt. Due to the lack of space, we omitted those parts that relies on other grammars, in particular: The EVENTDECL clause allows definition of events as first class citizens; it relies on the classes formulation typical of Manchester Syntax. An example of this is available in Listing 1.1. The CONSTRAINT clause allows the specification of filters; it relies on the SPARQL 1.1 grammar; an example of this is available in Listing 1.3. The user can specify time relations over semantic event declarations using the MATCH clause. Notably, the optional keyword **NAMED** works differently from SPARQL 1.1. It indicates which events the user is interested to select for the retrieval of the underlying RDF graph.

```
EventClause -> [NAMED] 'EVENT' EventIRI (EventDecl | PatternDecl)
EventDecl   ->   Follows Manchester Syntax Grammar^a
PatternDecl -> 'WHEN' PatternExpr [IFDecl] PatternExpr -> 'MATCH'
FollowedByExpr [WITHIN TimePeriod ] TimePeriod -> 'INTEGER' (ms | s
| m | h | d | w) FollowedByExpr -> OrExpr ((['NOT'] 'SEQ') OrExpr)*
OrExpr -> AndExpr ('OR' AndExpr)* AndExpr -> EveryOrNotExpr ( 'AND'
EveryOrNotExpr)* EveryOrNotExpr -> ['EVERY' | 'NOT' ] ( EventIRI
['AS' EventAltIri]
          | ( PatternExpr ) )*
IFDecl -> IF '{' 'EVENT' (EventIRI | Var) FilterExpr '}'
FilterExpr -> '{' ( BGP | 'FILTER' Constraint )*'}'
Constraint -> Follows the SPARQL 1.1 Grammar^b
```

Listing 1.4. Ontology-Based Event Processing Language Grammar. [a]https://www.w3.org/TR/owl2-manchester-syntax/#description. [b]https://www.w3.org/TR/rdf-sparql-query/\#rConstraint.

5   Ontology-Based Event Processing Architecture

In this section, we describe a system architecture for an OBEP system that supports the DELP syntax.

Figure 1 shows three different layers, each of which addresses a specific part of the processing to go from RDF Streams to the results of a DELP query. As anticipated in Sect. 4.3, we assume incoming events as a pair (G, t) where G is an RDF Graph and t is a timestamp (RDF Stream in Fig. 1).

Building on this assumption, Layer (a) is responsible for inferring high level concepts by applying reasoning over the incoming events; Layer (b) is responsible for identifying and extracting, from the underlying RDF graph, those properties that are relevant for filtering and joining, as specified in the query; last, but not least, Layer (c) applies event processing over the abstracted events as well as filtering and joining using the extracted properties. In the following paragraphs, each layer is described in detail.

To better understand how each layer behaves, we continue our running example. We want to capture the temporal relation between abnormal temperature and smoke in order to detect fire, but we need to ensure that the smoke detection and the abnormal temperature measure belong to the same room. In Listing 1.5, this requirements are translated into a time relation and a join condition: the variable ?v is used for the AbnormalTemperaturEvent and the SmokeDetectionEvent.

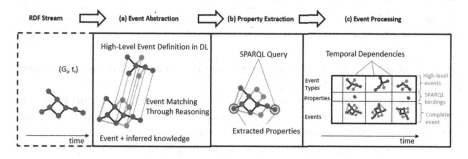

Fig. 1. Overview of the ontology-based event processing architecture

```
NAMED EVENT :FireEvent {
  MATCH :AbnormalTemperaturEvent -> :SmokeDetectionEvent WITHIN (5m)
  IF {
      EVENT :AbnormalTemperaturEvent { ?tmpSnsLoc a :Location.
                        ?tmpSnsLoc :hasValue ?v}
      EVENT :SmokeDetectionEvent { ?smkSnsLoc a :Location.
                        ?smkSnsLoc :hasValue ?v}
  }
}
```

Listing 1.5. Event Declaration for fire, if the smoke and temperature are sensed in the same location.

The incoming RDF graphs are added to the ABox, processed by the reasoner, and then removed. This process is show in Fig. 1a. DL reasoning is utilized, together with ontological definition of events, to materialize the incoming RDF graphs. When the reasoner, after a realization step, infers one of the defined high level events, these are forwarded to the next layer that can perform event processing over high level abstractions.

DELP allows the specification of filters and joins over the defined events. However, performing joins or filters requires to compare the values of those variables expressed in the DELP query. Which means access to the underlying RDF graph of high level ontological concepts that DELP targets. An additional SPARQL-querying layer, shown in Fig. 1b, is added in order to reach the underlying RDF graph that the high level event definition implies and extract the variables required for joining or filtering.

The translation from DELP filters to SPARQL queries is again straight forward. Listing 1.6 shows one of the required queries for the property extraction of the SmokeDectectionEvent in our example. For joins, the variable value must be the same for all the events sharing a variable; filters should positively validate a given conditional expression (e.g. lower than a specified threshold). Once the query is executed, the variable bindings are added to the event as properties, maintaining the naming convention. If no properties need to be extracted and no additional filtering is required, this step can be omitted.

```
SELECT ?tmpSnsLoc ?v
WHERE { ?tmpSnsLoc a : Location ; : hasValue ?v }
```

Listing 1.6. Translated SPARQL query for the property extraction based on the definition in Listing 1.5 for the SmokeDetectionEvent

The last layer in our proposed architecture is responsible for the actual event processing; it corresponds to Fig. 1c.

In our example, SmokeDetectionEvent and AbnormalTemperatureEvent are matched. Figure 2 zooms in Fig. 1c and shows the structure of the events once they reach the event processing layer for our running example: (Fig. 2I) the materialized events that therefore contain both explicit data (Blue) and those which have been inferred (Green); (Fig. 2II) the previously extracted values for variables involved in filters or, in this very case, joins; (Fig. 2III) the high level event definition, represented as an RDF graph to maintain a coherent notation.

Assuming such a layered data structure, the pattern matching can be translated into a target CEP language that provides filtering and joining using a name-based notation such as EPL. Listing 1.7 shows an example of this translation related to the fire detection example.

```
select * from pattern
[every a=AbnormalTemperaturEvent -> b=SmokeDetectionEvent(v=a.v)
 where timer:within(5 min)]
```

Listing 1.7. Event Declaration for fire, translated to EPL

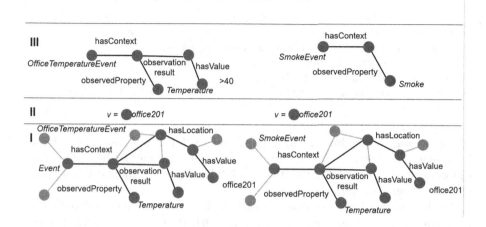

Fig. 2. Event processing over high level events. (Color figure online)

Building complex event structure is the goal of both CEP and SCEP systems. Therefore, it is worth discussing how complex events are provided to the user in case of positive pattern matching. At current stage, DELP does not include the specification of composed events explicitly. This is because it is hard to combine high level event description with their low level construction and we leave this as future work. Since event composition is crucial in event processing, we opt for a conservative solution and we define the complex event as the union of the underlying RDF graphs. The union is used since the event processor will only return values when the operator turned true. For example if E_1 has been detected and E_2 not, then E_1 OR E_2 will return true with E_2 as an empty collection.

Last but not least, we implemented an OBEP proof-of-concept system[6] containing the following technologies: the HermiT reasoner [7] for event abstraction in the first layer; Jena ARQ[7] for the property extraction of the underlying RDF graph in the second layer and the Esper engine[8] to perform the event processing on the high level events in the third layer.

6   Discussion and Conclusion

In this paper, we presented a first step towards ontology-based event processing. We designed an approach that contributes to the state-of-the-art of stream

---

[6] The code is part of the new version of MASSIF platform which is not yet available as open source. A stand alone version will be published at https://github.com/IBCNServices/OBEP.

[7] https://jena.apache.org/documentation/query/.

[8] http://www.espertech.com/.

Table 1. Differences and similarities between (S)CEP and OBEP approaches against Sect. 4 requirements. $\otimes$, i.e. SPARQL-like; $\star$, i.e. seq, and, or, not, every, within.

	R1.a	R1.b	R2	R3	R4 (filters)	R4 (joins)
EPL	Relational	/	$\star$	✓	✓	✓
EP-SPARQL [1]	RDF BGP	RDFS	seq, opt_seq eq_opt_seq	/	✓$^\otimes$	✓$^\otimes$
Taylor et al. [9]	OWL	Boh[a]	seq, or, and	✓	/	/
MASSIF [4]	DL Axioms	OWL 2 DL	/	/	✓	/
OBEP	DL Axioms	OWL 2 DL	$\star$	✓	✓$^\otimes$	✓$^\otimes$

[a]Riccardo: find.

reasoning with a requirement analysis; a syntax for Description Logic Event Processing, i.e. DELP; a three-layered architecture for an OBEP system that supports the proposed DELP syntax and fulfills our requirements; and a proof-of-concept implementation of a system.

Table 1 summarizes the differences and similarities between the related works mentioned in Sect. 2 and our approach for OBEP. This table highlights the novelty of the proposed system through the requirements that we presented in Sect. 4. Our approach combines semantic event declaration (R1.a) and event processing (R2). It also allows to compute temporal inference over the high-level concepts outputted by a deductive reasoning process. This is different from approaches that extend the ontological language to perform temporal inference, because they have to choose between either small entailments or intractability. DELP implements all the typical event processing operator (R3), while the other approaches focus on a subset. In particular, we include the *not*, which allows the definition of more expressive patterns. The final system complexity is composed by two layers, i.e. deductive reasoning and event processing. The second one is known to be polynomial in time, therefore the final complexity is bounded by the complexity of the ontological language used to describe the events.

In our future work, we will focus on the full language specification, i.e. full complexity description and the analysis under different DL fragments. We will investigate how to add the underlying definition of the events defined as RDF graphs. Integrating this in the language facilitates the creation of a more complete system that allows the processing of data on different levels. We aim at introducing explicit complex event construction semantics and also important time-aware operators, such as the ones of Allen's algebra. Finally, we plan to combine our approach with static knowledge for advanced inference and to thoroughly compare the performance of a prototype with state-of-the-art solutions, such as EP-SPARQL.

References

1. Anicic, D., Fodor, P., Rudolph, S., Stojanovic, N.: EP-SPARQL: a unified language for event processing and stream reasoning, pp. 635–644 (2011)
2. Barnaghi, P., Wang, W., Henson, C., Taylor, K.: Semantics for the Internet of Things: early progress and back to the future. Int. J. Semant. Web Inf. Syst. (IJSWIS) **8**, 1–21 (2012)

3. Bohlen, M.H., Busatto, R., Jensen, C.S.: Point-versus interval-based temporaldata models. In: Proceedings of the Fourteenth International Conference on Data Engineering, pp. 192–200 (1998)
4. Bonte, P., Ongenae, F., De Backere, F., Schaballie, J., Arndt, D., Verstichel, S., Mannens, E., Van de Walle, R., De Turck, F.: The MASSIF platform: a modular and semantic platform for the development of flexible IoT services. KAIS, 1–38 (2016)
5. Della Valle, E., Ceri, S., Harmelen, F.V., Fensel, D.: It's a streaming world! Reasoning upon rapidly changing information. IEEE Intell. Syst. **24**(6), 83–89 (2009)
6. Lutz, C., Wolter, F., Zakharyaschev, M.: Temporal description logics: a survey. In: 15th International Symposium on Temporal Representation and Reasoning, TIME 2008, Université du Québec à Montréal, pp. 3–14, 16–18 June 2008
7. Shearer, R., Motik, B., Horrocks, I.: HermiT: a highly-efficient owl reasoner. In: OWLED, vol. 432, p. 91 (2008)
8. Sirin, E., Bulka, B., Smith, M.: Terp: Syntax for owl-friendly SPARQL queries. In: Proceedings of the 7th International Workshop on OWL: Experiences and Directions (OWLED 2010), San Francisco, 21–22 June 2010
9. Taylor, K., Leidinger, L.: Ontology-driven complex event processing in heterogeneous sensor networks. In: Antoniou, G., Grobelnik, M., Simperl, E., Parsia, B., Plexousakis, D., Leenheer, P., Pan, J. (eds.) ESWC 2011. LNCS, vol. 6644, pp. 285–299. Springer, Heidelberg (2011). doi:10.1007/978-3-642-21064-8_20

Minimal Coverage for Ontology Signatures

David Geleta, Terry R. Payne[(✉)], and Valentina Tamma

Department of Computer Science, University of Liverpool, Liverpool, UK
{d.geleta,t.r.payne,v.tamma}@liverpool.ac.uk

Abstract. An ontology *signature* (set of entities) can express more than its constituent concept, role and individual names, since *rewriting* permits defined entities to be replaced by syntactically different, albeit semantically equivalent definitions. Identifying whether a given signature permits the definition of a particular entity is a well-understood problem, while determining the smallest (minimal) signature that covers a set of entities (i.e. a task signature) poses a challenge: the complete set of alternative definitions, or even just their signature, needs to be obtained, and all combinations of such definition signatures need to be explored, for each of the entities under consideration. In this paper, we present and empirically evaluate our novel approach for efficiently computing an approximation of *minimal signature cover sets*.

1 Introduction

An ontology provides a reference vocabulary for a given domain of interest, where the meaning of the terms in the vocabulary (entities) is *defined* inductively in terms of other entities [1]. Beth definability [2,6] is a well-known property from classical logic that has also been studied in the context of Description Logics (DLs); it relates the notion of *implicit definability* to the one of *explicit definability*, by stating that every implicitly defined concept is also explicitly definable [11]. In definitorially complete DLs ontologies (i.e. those modelled in a dialect where the *Beth definability* property holds), defined ontological entities can be rewritten into syntactically different, albeit semantically equivalent forms, thus it is possible to convey the meaning of an entity without using its actual name [11].

For example, in the ontology $\mathcal{O} = \{C \equiv A \sqcup B, C \equiv D, A \sqsubseteq \neg B, E \sqsubseteq \exists r.\top\}$, the concept C is defined explicitly, i.e. $C \equiv A \sqcup B$. Additionally, the concept A is implicitly defined in $\mathcal{O}$ by the set of general concept inclusions $\{C \equiv A \sqcup B, A \sqsubseteq \neg B\}$. Thus, A can be explicitly defined by the axiom $A \equiv C \sqcap \neg B$, where $\Sigma = \{C, B\}$ is a *definition signature* (DS) of A.

A concept or role can either be defined explicitly or implicitly in an ontology, or it could be *undefined* (and hence not rewritable, e.g. E, r). For instance, apart from the concept E, Σ *provides coverage* for all concept names of $\mathcal{O}$, as in addition to the asserted entities, both A and D are definable by an axiom whose signature is in Σ. Therefore, a signature potentially enables the expression of not only its asserted concept, role and individual names, but also of those defined entities whose definition is permitted with the given signature.

© Springer International Publishing AG 2017
M. Dragoni et al. (Eds.): OWLED-ORE 2016, LNCS 10161, pp. 128–140, 2017.
DOI: 10.1007/978-3-319-54627-8_10

Semantic interoperability between individually designed ontologies is typically hindered by heterogeneity, as distinct ontologies typically differ in their vocabularies and in the meaning they associate with particular entities [3]. *Ontology matching (alignment)* resolves heterogeneity between different ontologies by producing an alignment, i.e. a set of correspondences that describe relationships between semantically related entities of distinct ontologies. *Ontology alignment negotiation* has become an established and active research area that is concerned with supporting opportunistic communication within open environments [9,10]. In order to be able to carry out meaningful communication, different systems must cooperatively establish a mutually acceptable alignment, whilst adhering to internal preferences, without compromising confidential knowledge, and prevent alignment-based conservativity violations to occur [8]. The mutual alignment emerges as a result of a bilateral negotiation between two systems, whereby it is clearly beneficial to *minimise* the considered correspondences (i.e. the aligned part of an ontology signature), in order to reduce the overall cost of the process, and to support privacy and conservativity constraints of the interacting parties.

Van Harmelen and colleagues [12] have shown that semantic interoperability tasks can be characterised (amongst other parameters) by their signature, thus the prerequisite for performing a knowledge-based task is that the tasks's signature must be *covered* by the terms available to the party that performs the given task. In order to determine whether a given *task signature* (e.g. a query signature) is covered by a given *ontology signature*, each entity of the task signature must be individually examined; an entity is covered either if it appears in the ontology signature, or if it is rewritable using only the members of the ontology signature. Although task coverage is trivial to establish, determining the minimal signature that covers a given task signature poses a challenge, as the complete set of rewriting forms (definitions) needs to be known, and all combinations of such definition signatures (the sets of entities that permit the rewriting of a defined entity) are required to be explored, for each entity in question.

In our previous work, we have presented a pragmatic approach to computing the complete set of Definition Signatures for a given ontology [4]. In this paper we introduce the signature coverage problem and a novel algorithm that can efficiently compute an approximation of the smallest set of entities that covers a given task signature, starting from a previously obtained complete set of DSs. The paper casts the signature coverage problem as a set coverage problem, which is in Sect. 2, whilst Sect. 3 introduces the signature coverage problem. Section 4 presents our approach to approximate minimal cover sets. The approach is empirically evaluated in Sect. 5, that describes the experimental framework and the results. Finally, Sect. 6 outlines future work and concludes the paper.

2   Background

In this section we review two classical problems that address the issue of *coverage*: the set coverage problem and the minimal functional dependency cover. The *set coverage problem* (or minimal set cover problem) is a classic problem in combinatorics, complexity theory and computer science, in general [14]. Let $\mathcal{U}$ be a set of elements (referred to as the *universe*) and $\mathcal{S}$ a collection of subsets of $\mathcal{U}$, whose union equals the universe. The set cover problem identifies the smallest, *minimal* sub-collection $\mathcal{C} \subseteq \mathcal{S}$, called the *cover set*, such that the union of sets in $\mathcal{C}$ covers $\mathcal{U}$ $(\forall x\{x \in \mathcal{U}|x \in \mathcal{C}\})$.

Example 1 (Minimal set cover problem). Consider the set $\mathcal{U} = \{1, 2, 3, 4, 5\}$, and let $\mathcal{S}$ be the collection $\{S_1, S_2, S_3, S_4\}$, where $S_1 = \{1, 2, 3\}$, $S_2 = \{1, 2\}$, $S_3 = \{3, 4\}$ and $S_4 = \{4, 5\}$. The union of subsets of $\mathcal{S}$ contains all members of $\mathcal{U}$, thus $\mathcal{U}$ can be covered by an $S' \subseteq \mathcal{S}$. Although there are a number of possible solutions, there is only one minimal cover set, $\{S_1, S_4\}$.

Finding the minimal cover set is an NP-complete problem, however, there is a greedy algorithm that is able to find *approximations* (i.e. not necessarily minimal, but small cover sets) in polynomial time [14]. In the *weighted set cover* problem, each set $S_i \in \mathcal{S}$ is assigned a *weight* $w(\mathcal{S}) \geq 0$, and in this case, the goal is to find a cover set $\mathcal{C}$ with the minimal total weight $\sum_{S \in \mathcal{C}} w(S)$ (where the weight of a set does not correspond to its cardinality but emerges from the particular context where the set is used).

Another classical problem that has influenced the approach presented in this paper comes from relational database theory, and concerns finding the minimal functional dependency cover. A *functional dependency (FD)* is a constraint between two sets of attributes in a relation from a database [13]. For instance, given two attribute sets X and Y, then the FD $X \rightarrow Y$ means that the values of the attribute set Y are determined by the values of X, or in other words, two tuples in a database sharing the same values of X would also share the same values for Y. The *closure* of a set of attributes X with respect to a set of FDs $\mathcal{F}$ is the set X^+ of all attributes that are functionally determined by X using the closure of $\mathcal{F}$, denoted by $\mathcal{F}^+$. Before computing the closure, a set of FDs $\mathcal{F}$ is usually *normalised* by exhaustively applying inference rules (e.g. reflexivity, transitivity and augmentation) [13], as illustrated by the following example:

Example 2 (FD set attribute closures). Let us consider a set of FDs $\mathcal{F}$ such that

$$\mathcal{F} = \{(1)\ A \rightarrow B,\ (2)\ C \rightarrow E,\ (3)\ E \rightarrow F,\ (4)\ A, C \rightarrow D\}$$

$\mathcal{F}$ is already normalised (in third normal form), i.e. each FD contains exactly one attribute on the right-hand side (RHS), and each FDs' left-hand side (LHS) is irreducible. The closure of all attributes in $\mathcal{F}$ is given as follows:

1. $A^+ : A, B$ (A by reflexivity, B by (1))
2. $B^+ : B$ (B by reflexivity)
3. $C^+ : C, E, F$ (C by reflexivity, E by (2), F by transitivity and (2, 3))
4. $D^+ : D$ (D by reflexivity)
5. $F^+ : F$ (F by reflexivity)
6. $(A, C)^+ : A, B, C, D, E, F$ (A, C by reflexivity, B by (1), D by (4), E by (2), F by transitivity and (2, 3))

The closure of a set of attributes with respect to a set of FDs allows us to determine the minimal cover for a set of functional dependencies. A set of functional dependencies $\mathcal{F}$ covers another set of FDs $\mathcal{G}$ if every functional dependency in $\mathcal{G}$ can be inferred from $\mathcal{F}$, i.e. if $\mathcal{G}^+ \subseteq \mathcal{F}^+$. $\mathcal{F}$ is a *minimal cover* of $\mathcal{G}$ if $\mathcal{F}$ is the smallest set of functional dependencies that covers $\mathcal{G}$. It can be proven that every set of functional dependencies has a minimal cover, however this is not unique, as there may be more than one minimal cover.

3   The Signature Coverage Problem

The signature coverage problem determines whether a given *task signature* ($\mathcal{S}$) is covered by a *restricted signature* ($\mathcal{R}$), where both are subsets of the same vocabulary, i.e. the ontology signature ($\mathcal{R}, \mathcal{S} \subseteq \mathsf{Sig}(\mathcal{O})$). The restricted signature $\mathcal{R}$ could be obtained by considering only those entities in $\mathcal{O}$) that are mapped in an alignment. A task signature is said to be *covered* if all of its constituent entities are covered; by default, i.e. considering *explicit coverage*, this requires that $\mathcal{S} \subseteq \mathcal{R}$. Beth definability supports *implicit coverage*, where a task signature entity is replaceable by a definition axiom, if, given its signature Σ, $\Sigma \subseteq \mathcal{R}$. In order to achieve implicit coverage, the complete set of DSs needs to be obtained, where the number of possible rewritings (thus the number of unique DSs) of a defined concept or role is potentially exponential in the size of the ontology. Definition signatures may contain redundant elements, and could be as large as the ontology signature, thus we introduce the notion of signature minimality, that aims at minimising the size of a signature, by eliminating superfluous entities.

Definition 1 (Minimal Definition Signature (MDS)). *A signature Σ is a minimal definition signature of a defined entity e under a TBox $\mathcal{T}$, if none of its proper subsets are definition signatures of e.*

In previous work we have presented a pragmatic approach to compute in practice all Minimal Definition Signatures (MDSs). The approach focusses on the feasible defined entities (where given pre-defined complexity bounds all MDSs are computable) described using a DL language for which the Beth definability property holds [4]. Individual names can only be covered *explicitly* by an asserted entity, however, definable signature entities (concepts and roles) can also be covered *implicitly* as follows:

Definition 2 (Explicitly or implicitly covered entity). *Given an ontology $\mathcal{O}$, a task signature $\mathcal{S}$, and a restricted signature $\mathcal{R}$ such that $\mathcal{S}, \mathcal{R} \subseteq \mathsf{Sig}(\mathcal{O})$, an entity $e \in \mathcal{S}$ is covered explicitly by $\mathcal{R}$ iff $e \in \mathcal{R}$; or covered implicitly by $\mathcal{R}$, if there exists a DS Σ^e, such that $\Sigma \subseteq \mathcal{R}$; otherwise e is uncovered.*

A defined concept or role can simultaneously be covered explicitly and implicitly, thus a task signature entity $e \in \mathcal{S}$ may assume one of the four different *coverage status* w.r.t. a restricted signature $\mathcal{R}$, as shown in Table 1. Determining whether a given task signature is coverable by a particular, restricted signature is the

Table 1. Entity coverage status

COVERABILITY			REQUIRED COVERAGE
uncoverable		$e \notin \mathcal{R} \wedge \Sigma \nsubseteq \mathcal{R}$	explicit coverage: $e \in_? \mathcal{R}$
coverable	explicitly only	$e \in \mathcal{R} \wedge \Sigma \nsubseteq \mathcal{R}$	
	explicitly and implicitly	$e \in \mathcal{R} \wedge \Sigma \subseteq \mathcal{R}$	explicit and implicit coverage: $e \in_? \mathcal{R}^+$
	implicitly only	$e \notin \mathcal{R} \wedge \Sigma \subseteq \mathcal{R}$	

trivial process of identifying the coverage status of each task signature entity; i.e. for each entity $e \in \mathcal{S}$, we search for an MDS Σ^e such that $\Sigma^e \subseteq \mathcal{R}$. The set of entities covering all members of a task signature $\mathcal{S}$ is defined as follows:

Definition 3 (Cover set). *Given an ontology $\mathcal{O}$, a task signature $\mathcal{S}$, and a restricted signature $\mathcal{R}$ such that $\mathcal{S}, \mathcal{R} \subseteq \mathsf{Sig}(\mathcal{O})$, $\mathcal{C}$ is a cover set of $\mathcal{S}$ with respect to $\mathcal{R}$, if and only if $\mathcal{C} \subseteq \mathcal{R}$ and $\forall e\{e \in \mathcal{S} | e \in \mathcal{C} \vee \exists \Sigma^e | \Sigma^e \subseteq \mathcal{C}\}$.*

In other words, a cover set is a DS of all entities of the task signature. As an ontology signature can cover more than its constituent entities, due to the fact that a given signature may permits some defined entities to be implicitly covered, we adopt the notion of *closure* from FD computation, to provide a representation which describes *the set of all entities covered* by a given signature:

Definition 4 (Signature closure). *Given an ontology $\mathcal{O}$, and a signature $\mathcal{X}$ such that $\mathcal{X} \subseteq \mathsf{Sig}(\mathcal{O})$, the signature closure $\mathcal{X}^+$ contains all explicitly and implicitly covered entities of $\mathsf{Sig}(\mathcal{O})$ by $\mathcal{X}$, i.e. $\forall e\{e \in \mathsf{Sig}(\mathcal{O}) | e \in \mathcal{X} \vee \exists \Sigma^e | \Sigma^e \subseteq \mathcal{X}\}$.*

This permits a more succinct definition of coverage: a task signature $\mathcal{S}$ is covered by a set $\mathcal{C}$ iff $\mathcal{S} \subseteq \mathcal{C}^+$. Once it has been established, that the $\mathcal{S}$ is coverable by $\mathcal{R}$, the problem is to identify the *smallest subset* $\mathcal{C} \subseteq \mathcal{R}$ which covers $\mathcal{S}$. This is referred to as the minimal cover set and defined as follows:

Definition 5 (Minimal cover set). *Given an ontology $\mathcal{O}$, a task signature $\mathcal{S}$, a restricted signature $\mathcal{R}$ such that $\mathcal{S}, \mathcal{R} \subseteq \mathsf{Sig}(\mathcal{O})$, and the set $\mathcal{C}$ which covers $\mathcal{S}$ with respect to $\mathcal{R}$, $\mathcal{C}$ is minimal if and only if there is no other cover set $\mathcal{C}' \subseteq \mathcal{R}$ such that $|\mathcal{C}'| < |\mathcal{C}|$.*

There can be more than one, unique minimal cover set, i.e. two sets with the same cardinality whose complement is not an empty set. Finding a minimal cover set is an non-polynomial problem, because it requires all cover sets to be identified, by exhaustively testing each subset of the power set of the ontology signature, in order to find all valid covers and select the one with the minimum cardinality. *Approximation algorithms* are commonly used for problems with NP time complexity, such as the set cover problem, to provide sub-optimal solutions in polynomial-time. The *greedy algorithm design* is one of the standard techniques for approximation algorithms [14]. The next example illustrates the signature cover problem, and outlines a cover set approximation approach.

Example 3 (Signature Coverage). Let $\mathcal{O}$ be an ontology, $\mathcal{S}$ a task signature, $\mathcal{R}$ a restricted signature, and M the complete set of MDSs of each defined entity of $\mathcal{S}$ (an MDS $m \in M$ is represented as the tuple $\langle e, \Sigma \rangle$, where e is the defined concept or role name, and Σ denotes the MDS), where $\mathcal{S}, \mathcal{R} \subseteq \mathsf{Sig}(\mathcal{O})$, such that

- $\mathsf{Sig}(\mathcal{O}) = \{\mathsf{A}, \mathsf{B}, \mathsf{C}, \mathsf{D}, \mathsf{E}, \mathsf{F}, \mathsf{r}, \mathsf{s}, \mathsf{q}\}$
- $M = \{\langle \mathsf{C}, \{\mathsf{A}, \mathsf{B}\}\rangle, \langle \mathsf{C}, \{\mathsf{E}, \mathsf{r}\}\rangle, \langle \mathsf{C}, \{\mathsf{q}\}\rangle, \langle \mathsf{B}, \{\mathsf{D}\}\rangle, \langle \mathsf{D}, \{\mathsf{B}\}\rangle, \langle \mathsf{s}, \{\mathsf{r}\}\rangle\}$
- $\mathcal{S} = \{\mathsf{B}, \mathsf{C}, \mathsf{D}, \mathsf{E}, \mathsf{s}, \mathsf{q}\}$
- $\mathcal{R} = \{\mathsf{A}, \mathsf{B}, \mathsf{C}, \mathsf{D}, \mathsf{E}, \mathsf{r}, \mathsf{q}\}$

Without accounting for definability, i.e. by only considering explicit coverage, the $\mathcal{R}$ does not cover the task signature because $\mathcal{S} \backslash \mathcal{R} \neq \emptyset$. However, considering implicit coverage shows that the closure of the restricted signature is $\mathcal{R}^{+} = \{\mathsf{A}, \mathsf{B}, \mathsf{C}, \mathsf{D}, \mathsf{E}, \mathsf{r}, \mathsf{s}, \mathsf{q}\}$, thus $\mathcal{S}$ can be covered by $\mathcal{R}$, because $\mathcal{S} \subseteq \mathcal{R}^{+}$. Following a naive, greedy approach, one may select those entities that appear both in $\mathcal{S}$ and $\mathcal{R}$ as these entities can be covered explicitly, i.e. $C_1 = \mathcal{S} \cap \mathcal{R} = \{\mathsf{C}, \mathsf{D}, \mathsf{E}, \mathsf{q}\}$; then attempt to cover the remaining task signature entities, by adding a corresponding MDSs for each uncovered task signature entity; in this case covering s by adding r to C_1, as there is an MDS $\langle \mathsf{s}, \{\mathsf{r}\}\rangle$ thus s is implicitly definable by the signature $\{\mathsf{r}\}$. As a result $C_1 = \{\mathsf{B}, \mathsf{C}, \mathsf{D}, \mathsf{E}, \mathsf{r}, \mathsf{q}\}$ covers $\mathcal{S}$. However, the smallest cover set is $C_2 = \{\mathsf{B}, \mathsf{E}, \mathsf{r}, \mathsf{q}\}$, because $C_2^{+} = \{\mathsf{B}, \mathsf{C}, \mathsf{D}, \mathsf{E}, \mathsf{r}, \mathsf{s}, \mathsf{q}\}$, $\mathcal{S} \subseteq C_2^{+}$, and $|C_1| > |C_2|$.

In Example 3, a naive, greedy approach (*Greedy #1*) has produced a non-minimal cover set C_1, which was an approximation of the minimal cover C_2. The cover set C_1 can be improved by removing redundant entities (resulting in the set $C_1' = C_2$), i.e. producing a non-redundant cover set:

Definition 6 (Non-redundant cover set). *Given an ontology $\mathcal{O}$, a task signature $\mathcal{S}$, a restricted signature $\mathcal{R}$ such that $\mathcal{S}, \mathcal{R} \subseteq \mathsf{Sig}(\mathcal{O})$, and C which covers $\mathcal{S}$ w.r.t. $\mathcal{R}$, C is non-redundant iff none of its proper subsets cover $\mathcal{S}$.*

Non-redundant cover sets are typically small, however, as there can be more than one non-redundant cover set with different cardinality, a non-redundant cover set is not necessarily the minimal cover set. It is worth noting that every minimal cover set is also a non-redundant set.

4   Approximating Minimal Cover Sets

In order to tackle the exponential time complexity of the minimal signature cover problem, we introduce a greedy, approximation algorithm that provides a sub-optimal solution in polynomial-time; where the resulting cover set is always non-redundant. The basic idea behind the approach is that by starting from an empty set, the cover set is built up incrementally until all task signature members are covered, however, instead of selecting individual entities from the restricted signature, at each iteration the approach selects an entity set. The entity sets that are being considered are MDSs, because individual entities typically only provide explicit coverage (for themselves or their synonyms), while MDSs can cover defined entities implicitly (in addition to explicitly covering all those task signature entities that appear in the MDS as well). The selection is made by assigning a *cost* and *value* score to each MDS, and then picking the MDS which provides the maximum value and the minimum cost with respect to the task signature and the incomplete cover set, prioritising on the value score. The cost quantifies the number of entities required to be added to the cover set (i.e. the set difference of the cover set and the particular MDS), while the value represents the number of entities that the given signature covers (an MDS can be a DS for more than one defined entity, thus it can cover several task signature entities). In case there are more than one MDSs with the same cost and value, a random MDS is selected.

In order to evaluate the *actual value* of a given MDS, i.e. the set of all entities of the task signature that the MDS covers either explicitly or implicitly, similarly to FDs, its *closure* needs to be identified, thus we represent MDSs in the form of FDs to facilitate this notion. There is a strong resemblance between the concept of an FD and an MDS, meaning that an MDS can be thought of as a dependency between entities of an ontology, where the relation between the signature of the LHS and the entity on the RHS is implicit definability. For example, the MDS $\Sigma^{\mathsf{C}} = \{\mathsf{A}, \mathsf{B}\}$ which defines concept C using entities $\{\mathsf{A}, \mathsf{B}\}$ may be represented as $m : (\mathsf{A}, \mathsf{B} \to \mathsf{C})$; such an MDS is referred to as an fMDS, and defined as follows:

Definition 7 (fMDS). *Given a defined entity e, and its minimal definition signature Σ, where $e \in \mathsf{Sig}(\mathcal{O})$ and $\Sigma \subseteq \mathsf{Sig}(\mathcal{O})$, the corresponding fMDS is the function $m : (\Sigma \to e)$, which, given the entity set Σ, implicitly covers e.*

Analogously to functional dependencies, the closure of an fMDS is computed from the *set of all fMDSs*, by identifying all relevant MDSs:

Definition 8 (fMDS closure). *Given an fMDS $m_i : (\Sigma \to e)$, and a set of fMDSs M where $m_i \in M$, the closure of m_i is the function $m_i^+ : (\Sigma \to E)$ such that*

$$E = \Sigma \cup \{e\} \cup \left(\bigcup \forall m_j^{+RHS} \{m_j \in M | m_j^{+LHS} \subseteq m_i^{+LHS}\} \right)$$

where m^{+LHS} denotes the signature Σ, and m^{+RHS} refers to signature E.

The closure of a set of fMDSs M is the set M^+, where each $m_i^+ \in M^+$ is the closure of the corresponding $m_i \in M$; this is illustrated by the next example:

Example 4 (fMDS set closure). Let M be a set of fMDS such that

- $M = \{m_1 : (\mathsf{A}, \mathsf{B} \to \mathsf{C}), m_2 : (\mathsf{B} \to \mathsf{D})\}$
- $M^+ = \{m_1^+ : (\mathsf{A}, \mathsf{B} \to \mathsf{A}, \mathsf{B}, \mathsf{C}, \mathsf{D}), m_2^+ : (\mathsf{B} \to \mathsf{B}, \mathsf{D})\}$

The closure of fMDSs (M^+) is computed as follows:

1. m_1^{+LHS} in addition to implicitly covering concept C, also explicitly covers concepts A, B, hence $m_1^{+RHS} = m_1^{RHS} \cup \{\mathsf{A}, \mathsf{B}\}$;
2. m_1^+ implicitly covers D as $m_2^{+LHS} \subseteq m_1^{+RHS}$ thus $\mathsf{D} \in m_1^{+LHS}$;

Algorithm 1. COMPUTEMINIMALSIGNATURECOVER($\mathcal{O}, \mathcal{R}, \mathcal{S}, M$)

Input : $\mathcal{O}$: ontology; $\mathcal{S}$: task signature; $\mathcal{R}$: restricted signature;
M: the complete set of MDSs of each defined entity $e \in \mathcal{O}$

Output: $\mathcal{C}$: non-redundant cover set of $\mathcal{S}$ w.r.t. $\mathcal{R}$ if and only if $\mathcal{S} \subseteq \mathcal{R}^+$

1   $\mathcal{C} \leftarrow \mathcal{C} \cup \forall e\{e \in \mathcal{S} \cap \mathcal{R} | e \notin m_i^{RHS} | m_i \in M\}$

2   $M^+ \leftarrow$ Initialise(M)

3   $\mathcal{C}^+ \leftarrow$ ComputeSignatureClosure($\mathcal{C}, M^+$)

4   $\overline{M^+} \leftarrow M^+ \backslash \forall m_i \{m_i \in M^+ | m_i^{LHS} \subseteq \mathcal{C}^+\}$

5   **while** ($\mathcal{S} \backslash \mathcal{C}^+) \neq \emptyset$ **do**

6      $\mathcal{V} \leftarrow$ ComputeValueCostVector($M^+, \mathcal{S}, \mathcal{C}^+$)

7      $m' \leftarrow$ select an $m \in \overline{M^+}$ according to $\mathcal{V}$, with max $v(m)$, and min $c(m)$

8      $\mathcal{C} \leftarrow \mathcal{C} \cup m'^{LHS}$

9      $\mathcal{C}^+ \leftarrow$ ComputeSignatureClosure($\mathcal{C}, M$)

10     $\overline{M^+} \leftarrow \overline{M^+} \backslash (\{m'\} \cup \forall m_i \{m_i \in \overline{M^+} | m_i^{LHS} \subseteq \mathcal{C}^+\})$

11 **end**

12 **return** $\mathcal{C}$

---

3. m_2 explicitly covers concepts B, hence $m_2^{+RHS} = m_2^{RHS} \cup \{\mathsf{B}\}$;
4. no more MDSs apply, thus the closure is complete.

The cost and value calculation of an fMDS is formalised as follows:

Definition 9 (fMDS value and cost). *Given an fMDS m, an ontology $\mathcal{O}$, a task signature $\mathcal{S}$, a cover set $\mathcal{C}$, and M the complete set of $fMDSs$ in $\mathcal{O}$, where $m, \mathcal{S} \subseteq \mathcal{O}$, the value and cost of m with respect to $\mathcal{S}$ and $\mathcal{C}$ is given by the value function $v(m) = |\mathcal{R} \backslash \mathcal{C}^+ \cap m^{RHS}|$, and the cost function $c(m) = |\mathcal{C}^+ \backslash m^{LHS}|$, where both $v(m)$ and $c(m)$ assign a natural number $i \in \mathbb{N}_0$ to m.*

Algorithm 1 formalises the approximation approach, which employs two sub-routines for computing the *closure of signatures*, and the *closure of fMDSs*. The algorithm first applies an *optimisation heuristic*, which reduces the search space by initialising cover $\mathcal{C}$ with the only explicitly coverable entities of $\mathcal{S}$ (line 1). Next M, which is used as the search space, is initialised with the complete set of fMDSs. In addition, the process generates an 'artificial' MDS and stores it in M^+, for each entity that can be covered both explicitly and implicitly. For instance, given a concept A, the generated fMDSs is $\mathsf{m} : (\mathsf{A} \rightarrow \mathsf{A})$, i.e. the entity can cover itself. By including such artificial fMDSs that do not originate from actual MDSs, the algorithm ensures that the search space is complete, i.e. for each $e \in \mathcal{S}$ the search space M^+ includes *all* possible ways of cover. The initialisation is concluded by computing the fMDS set closure (line 2). Before the process begins the search, $\mathcal{C}^+$, the cover closure is computed. This facilitates the *termination condition of the search* process (line 5) that halts the algorithm when $\mathcal{S}$ is covered (i.e. $\mathcal{S} \backslash \mathcal{C}^+ = \emptyset$). Then $\overline{M^+}$ is created as a copy of M^+, the former is the actual search space which is continuously pruned at each iteration (to optimise the process, by reducing the search space and subsequently the effort required to calculate the cost and value scores of fMDSs), while the latter is left intact for the purpose of computing signature closures during the search. $\overline{M^+}$ is pruned out by removing any fMDS whose value (and cost) w.r.t. $\mathcal{C}$ is zero (line 4). During the search (line 5–11), the value and cost of each fMDS in $m_i \in \overline{M^+}$ is evaluated w.r.t. the cover (line 6), then the best fMDS is selected (line 7) and the LHS of the fMDS (i.e. the MDS) is added to the cover (line 8). The cover is then reevaluated by updating its closure (line 9), finally $\overline{M^+}$ is pruned according to the updated cover set. These steps are repeated until $\mathcal{S}$ is covered, at which point the algorithm returns the completed cover.

The algorithm always finds a non-redundant cover set, this is ensured by the selection function, and the fact that the entities added to the cover are MDSs, i.e. already minimal entity sets that are required to cover an other entity. At the worst case, the process covers at least one entity at each iteration, thus the maximum number of steps performed by the algorithm is n, where $n = |\mathcal{S}|$. As both subroutines employed by this algorithm have polynomial time computational complexity, it holds that the overall complexity is polynomial in the size of the input as well. Moreover, as both of subroutines terminate, and the halting condition (line 5) suspends the main loop of Algorithm 1 when the cover set is complete, it follows that the Algorithm 1 also terminates. A more exhaustive description of the presented algorithms can be found in [5].

5   Empirical Evaluation

In this section, we empirically determine how effective our approximation approach is in finding cover sets. The evaluation tested the *hypothesis* that the presented approach (*Greedy #2*), by considering both explicit and implicit coverage, produces a cover set that, although not minimal, is still significantly smaller than cover sets obtained by only explicit coverage. Thus approximations of minimal cover sets are typically smaller than explicit covers, if the given task signature contains defined entities w.r.t. a restricted signature (clearly, for a task signature which lacks defined entities only explicit coverage is possible).

The *evaluation corpus* was assembled from several semantically rich OWL ontologies that are commonly used for empirical evaluation in the ontology matching and alignment negotiation literature. We have selected 7 small ontologies (average 114.57 concepts and roles, and 307.57 axioms per ontology) from the *Conference* dataset[1], which describes the conference organisation domain; and 2 large (average 13397.00 entities, and 14663.00 axioms) ontologies from the *Large biomedical* dataset[2]. For every concept and role in each ontology, we have pre-computed the definability status and the complete set of MDSs. Table 2 presents a summary of the corpus, showing the DL expressivity, the number of logical axioms, number of concept and roles ($\mathcal{C} \cup \mathcal{R}$) in the ontology signature, the ratio of defined entities to all entities ($Def\%$), and the average number of different MDSs per defined entity ($\mathcal{M}$). Both datasets contain ontologies with

---

[1] http://oaei.ontologymatching.org/2014/conference/index.html.
[2] http://oaei.ontologymatching.org/2014/largebio/index.html.

Table 2. Comparing ideal covers with approximations produced by approach #2, for covering the entire ontology signature.

Ontology	$\mathcal{DL}$ expressivity	Axioms	$(\mathcal{C} \cup \mathcal{R})$	$Def\%$	$\mathcal{M}$	Ideal Cover	Greedy #2	
						cov	cov	Time
Conference corpus								
cmt	$\mathcal{ALCIN}(\mathcal{D})$	226	88	50.00%	1.09	50.00%	72.73%	3.62 ms
conference	$\mathcal{ALCHIF}(\mathcal{D})$	285	123	57.72%	1.54	42.28%	70.73%	11.79 ms
confOf	$\mathcal{SIN}(\mathcal{D})$	196	74	12.16%	3.33	87.84%	89.19%	0.43 ms
edas	$\mathcal{ALCOIN}(\mathcal{D})$	739	153	26.14%	2.80	73.86%	86.27%	8.70 ms
ekaw	$\mathcal{SHIN}$	233	106	28.30%	1.00	71.70%	85.85%	2.30 ms
iasted	$\mathcal{ALCIN}(\mathcal{D})$	358	181	17.68%	1.75	82.32%	87.85%	3.18 ms
sigkdd	$\mathcal{ALEI}(\mathcal{D})$	116	77	25.97%	1.55	74.03%	81.82%	1.12 ms
AVG.		**307.57**	**114.57**	**31.14%**	**1.87**	**68.86%**	**82.06%**	**4.45 ms**
LargeBio corpus								
NCI_fma	$\mathcal{ALC}$	9083	6551	29.98%	1.32	70.02%	70.02%	3.09 s
SNOMED_fma	$\mathcal{ALER}$	20243	13430	21.47%	1.09	78.54%	78.56%	8.64 s
AVG.		**14663.00**	**13397.00**	**25.72%**	**1.16**	**74.28%**	**74.29%**	**5.86 s**

varying level of definability, as shown by the ratio of defined ontology signature entities and the number of different MDSs per entity.

The _experimental framework_ was implemented in Java; the OWL API [7] was used for ontology manipulation; entity definability status, and MDSs were computed using the OntoDef API [4]. In all experiments, for each task signature, we have computed cover sets by using the approximation approach. We have only considered _coverable_ task signatures (i.e. $\mathcal{S} \subseteq \mathcal{R}^+$), thus in all cases, the restricted signature $\mathcal{R}$ was equivalent to the T-Box signature, while the task signature $\mathcal{S}$ was allocated several differently sized T-Box signature subsets (i.e. $\mathcal{R} = \mathsf{Sig}(\mathcal{T})$, and $\mathcal{S} \subseteq \mathsf{Sig}(\mathcal{T})$). Varying the composition of only one of the two signatures simplified both the experiment conduct and the result analysis process. Experiments were conducted with 8 GB maximum memory allocated for the JVM, running on a machine equipped with 16 GB RAM and a 4-core 2.9 GHz 64-bit processor architecture.

Experiment 1: Full signature. This experiment compares the size of the approximated covers to the _ideal cover_. The only cover which is potentially minimal and can be computed efficiently, is obtainable when the task requires the entire signature of an ontology to be covered ($\mathcal{S} = \mathsf{Sig}(\mathcal{O})$). This special case provides the opportunity to evaluate the difference between an actual, and an approximated minimal cover set. The ideal cover is obtained by removing all non-redundant, defined entities from the ontology signature. Considering non-redundancy in role removal is necessary in order to avoid removing those entities that are both defined, and provide the only DS to another entity. For example, the axiom r ≡ s implies that both roles r and s are defined, however removing both entities from the ideal cover would make them both uncoverable. Table 2 presents the experiment results, showing the size of the cover set in relation to the an explicit cover (which is always equivalent to the task signature hence $cov = \frac{|C|}{|S|}$), while the RHS partition present the result obtained by the greedy algorithm, in terms of the cover to task signature size ratio, and computation time (wall time). The approach achieves considerable reduction in all ontologies, on average the approximated cover is 82.06% of the explicit cover in the Conference, and 68.86% in the LargeBio corpus (which only falls short by 0.01% from the optimal solution, i.e. the ideal cover).

Experiment 2: Varying signatures. The second experiment was carried out using different task signature sizes, in order to assess the reduction provided by a minimal cover in comparison with the baseline (explicit cover), and to evaluate the computation time on a wider scale of possible tasks sizes. The experiment included 20 test cases for each ontology, where the task to ontology signature ratio ranged between 100% and 5%. Due to the fact that cover computation includes a non-deterministic part, where a random choice is made to select an MDS from a set of equally good options (MDSs with the same value and cost scores), each test case was repeated over 100 times (we have tested several different repetition counts, and by comparing their relative standard deviation established that 100 repetitions were sufficient for both datasets). Figure 1(a, b) presents the *cover set cardinality* results, where the y-axis represents the approximated minimal cover to task

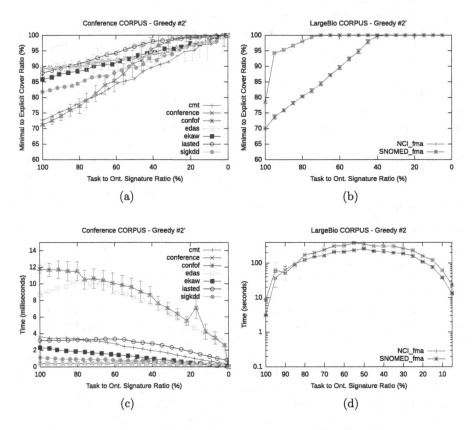

Fig. 1. Cover set *sizes* (*a, b*) and *computation times* (*c, d*) in the corpus.

signature ratio ($\frac{|\mathcal{C}|}{|\mathcal{S}|}$), and the x-axis shows the task signature to ontology signature ratio ($\frac{|\mathcal{S}|}{|\mathsf{Sig}(\mathcal{O})|}$); for brevity, error bars are only shown for the ontologies with the highest and lowest covers. The approach achieves varying level of reduction in all ontologies, where the efficiency decreases with the size of the task signature (due to the lower probability of the signature containing defined entities). Figure 1(c, d) presents the *computation time* (wall time) results, where the y-axis shows the time (either in milliseconds for the Conference, or in seconds for the LargeBio corpus), and the x-axis shows the task to ontology signature size ratio. The results suggest that the approach is in practice feasible for both small and large ontologies.

6   Conclusions and Future Work

In this paper, we have introduced and characterised the ontology signature coverage problem, which is a non-polynomial time problem that concerns whether an ontology signature can be covered by another signature, under a given ontology. Furthermore, we have presented and empirically evaluated a novel approach that, by exploiting the notion of Beth-definability in DL ontologies and using the pre-computed, complete set of different MDSs, provides a sub-optimal solution to the minimal signature cover problem. The evaluation has confirmed that, although the resulting covers are not necessarily minimal, the presented greedy approximation approach provides significant reduction in cover set size than only explicit coverage. The approach identifies a single cover, however, there can be more than one minimal cover set, hence as future work we will explore identifying multiple covers.

References

1. Baader, F.: The Description Logic Handbook: Theory, Implementation, and Applications. Cambridge University Press, Cambridge (2003)
2. Beth, E.W.: On Padoa's method in the theory of definition. Indagationes Mathematicae **15**, 330–339 (1953)
3. Euzenat, J., Shvaiko, P.: Ontology Matching, 2nd edn. Springer, Heidelberg (2013)
4. Geleta, D., Payne, T.R., Tamma, V.: An investigation of definability in ontology alignment. In: Blomqvist, E., Ciancarini, P., Poggi, F., Vitali, F. (eds.) EKAW 2016. LNCS (LNAI), vol. 10024, pp. 255–271. Springer, Cham (2016). doi:10.1007/978-3-319-49004-5_17
5. Geleta, D., Payne, T.R., Tamma, V.: Computing minimal signature coverage for description logic ontologies. Technical report, ULCS-16-004, University of Liverpool (2016)
6. Hoogland, E., et al.: Definability and interpolation: model-theoretic investigations. Institute for Logic, Language and Computation (2001)
7. Horridge, M., Bechhofer, S.: The OWL API: a Java API for OWL ontologies. Semant. Web **2**(1), 11–21 (2011)
8. Jiménez-Ruiz, E., Payne, T.R., Solimando, A., Tamma, V.: Avoiding alignment-based conservativity violations through dialogue. In: Proceedings of the OWLED, vol. 15 (2015)

9. Payne, T.R., Tamma, V.: Using preferences in negotiations over ontological correspondences. In: Chen, Q., Torroni, P., Villata, S., Hsu, J., Omicini, A. (eds.) PRIMA 2015. LNCS (LNAI), vol. 9387, pp. 319–334. Springer, Cham (2015). doi:10.1007/978-3-319-25524-8_20

10. Santos, G., Tamma, V., Payne, T.R., Grasso, F.: Dialogue based meaning negotiation. In: The 15th Workshop on Computational Models of Natural Argument (CMNA 2015) (2015)

11. Ten Cate, B., Franconi, E., Seylan, I.: Beth definability in expressive description logics. J. Artif. Intell. Res. (JAIR) **48**, 347–414 (2013)

12. Van Harmelen, F., Ten Teije, A., Wache, H.: Knowledge engineering rediscovered: towards reasoning patterns for the semantic web. In: Fensel, D. (ed.) Foundations for the Web of Information and Services, pp. 57–75. Springer, Heidelberg (2011)

13. Vardi, M.Y.: Fundamentals of dependency theory. IBM Thomas J. Watson Research Division (1985)

14. Vazirani, V.V.: Approximation Algorithms. Springer, Heidelberg (2013)

OWL API for iOS: Early Implementation and Results

Michele Ruta[✉], Floriano Scioscia, Eugenio Di Sciascio, and Ivano Bilenchi

Politecnico di Bari, via E. Orabona 4, 70125 Bari, Italy
{michele.ruta,floriano.scioscia,eugenio.disciascio}@poliba.it,
ivanobilenchi@gmail.com

Abstract. Semantic Web and Internet of Things are progressively converging, but the lack of reasoning tools for mobile devices on the iOS platform may hinder the progress of this vision. The paper presents an early redesign of OWL API for iOS. A partial port has been developed, effective enough to support mobile reasoning engines in a moderately expressive fragment of OWL 2. Both architecture and mobile-oriented optimization are sketched and preliminary performance results are discussed.

1 Introduction and Motivation

Semantic Web technologies are a key enabler of interoperability and intelligent information processing not only in the WWW, but also in the so-called Internet of Things (IoT). Application scenarios include supply chain management [5], (mobile) sensor networks [14], building automation [15] and more. The Semantic Web and the IoT paradigms are progressively overlapping in the *Semantic Web of Things* (SWoT) vision [14,17]. SWoT enables semantic-enhanced pervasive computing by associating informative fragments to multiple heterogeneous micro-devices in a given environment, each acting as a knowledge micro-repository. Rather than the batch processing of large ontologies and complex inferences prevalent in traditional Semantic Web scenarios, SWoT requires quick reasoning and query answering on sets of relatively elementary resources, in order to provide mobile agents with on-the-fly autonomous decision capabilities. The ever-increasing computing potentialities of mobile devices allow processing of rich and formally structured information without resorting to centralized nodes and support infrastructures. For a full accomplishment of this vision, reasoning engines and library interfaces are needed on the most relevant mobile device platforms.

iOS is the second largest mobile Operating System (OS) worldwide, with over 1 billion iPhone units sold (as of July 2016 [1]) as well as iPad and iPod Touch devices. While Android has a larger active device count, iOS has been more eagerly adopted in business [22]. Higher hardware and OS uniformity, a stricter security model [12], enterprise IT (Information Technology) department support tools and a stronger focus on usability are among the reasons. Business sectors

© Springer International Publishing AG 2017
M. Dragoni et al. (Eds.): OWLED-ORE 2016, LNCS 10161, pp. 141–152, 2017.
DOI: 10.1007/978-3-319-54627-8_11

ranging from healthcare to sales management and research exhibit a thriving market of iOS software solutions. Nevertheless, a full adoption of Semantic Web technologies has not been possible on iOS so far. A recent survey [11] found no Web Ontology Language (OWL) [21] reasoners implemented in Objective-C or Swift, the only two languages natively supported on iOS. In fact Java is by far the most popular implementation language for that. Several reasoners originally developed for Java Standard Edition have been ported to the Java-based Android platform so as to run on mobile devices [3]; likewise Java-based reasoning engines expressly designed for mobile devices also exist, including *mTableau* [19] and *Mini-ME* [18], which work on Java Micro Edition and Android, respectively. Similarly, all main OWL Knowledge Base (KB) management libraries are Java-oriented. Among them the *OWL API* [7] is the most adopted one. Java code requires a rewriting effort toward Objective-C or Swift in order to be adopted on iOS (whereas C/C++ list can be reused in Objective-C projects by writing proper wrappers).

The lack of iOS Semantic Web tools hampers the development of multi-platform semantic-enabled mobile applications to follow the rapid pace of the IoT (r)evolution, which may stifle the SWoT vision as a whole [6]. Although toolkits (such as *Oracle Mobile Application Framework*[1] and *Codename One*[2]) allow cross-platform mobile development in Java language and deployment to iOS devices, they are affected by various cost, efficiency and inconvenience issues. Automatic source transpilers from Java to Objective-C (such as *J2ObjC*[3]) also exist, but they are primarily intended to allow multi-platform projects to share as much business logic code as possible: transpiling existing software is significantly harder from a development point of view, especially considering that library dependencies must be recursively translated, or suitable alternatives need to be found or developed. Automatic translation is also not very flexible, as the core architecture of the source project cannot be altered without a considerable amount of work: this is an issue in this specific case, since significant architectural changes to the OWL API internals are desirable in order to ensure high performance (both in terms of time and memory) in SWoT scenarios.

In order to allow developing mobile reasoners for iOS, we present here the first results of porting the OWL API to iOS. This approach was preferred over writing a new application programming interface because the OWL API is a *de facto* standard for manipulating DL KBs and has a large user community. A functional subset of the OWL API was implemented, able to load and process KBs in an OWL 2 fragment corresponding to the $\mathcal{ALEN}$ Description Logic (DL) –with the addition of role hierarchies– in RDF/XML syntax. The ported library was written in Objective-C, to be used by both Objective-C and Swift applications. It runs on iOS and macOS without modification, as it does not use iOS-specific APIs. Experimental tests verified the correctness of the implementation and exhibit satisfactory results also in comparison with the original Java OWL API

---

[1] http://www.oracle.com/technetwork/developer-tools/maf/overview/index.html.

[2] https://www.codenameone.com/.

[3] http://j2objc.org.

on macOS. The library is released[4] as open source under the *Eclipse Public License* and can already support a future Mini-ME port for iOS.

The remainder of the paper is as follows: Sect. 2 provides background on the OWL API and porting strategies, while Sect. 3 describes the developed library; experimental results are in Sects. 4 and 5 closes the work.

2 Background

The OWL API [7] is the most commonly used front-end for OWL-based Knowledge Base Management Systems (KBMS) [3,11]. Other interfaces include Jena[5], Protégé-OWL API [8] and OWLlink [10]. The Jena library provides ontology manipulation APIs for Resource Description Framework (RDF) [16], RDF Schema (RDFS) [4] and OWL models, and an inference API to support reasoning and rule engines. The Protégé-OWL API [8] leverages Jena on OWL and is particularly effective for developing graphical applications. OWLlink [10] is a client/server protocol on top of HTTP for KB management and reasoning. The OWLlink API [13] implements OWLlink on top of the OWL API and therefore could be also ported to iOS.

The OWL API is a Java library defining a set of interfaces to manipulate OWL 2 KBs. It supports loading and saving in several syntaxes, including RDF/XML, Turtle, the Manchester Syntax and more. The implemented model gives an abstract representation of concept, property, individual and axiom types in OWL 2 through four interface hierarchies, all having `OWLObject` as a common ancestor. The model interfaces do not depend on any particular concrete syntax. The `OWLOntologyManager` interface allows creating, loading, changing and saving KBs, alleviating the burden of choosing the appropriate parsers and renderers. Finally, `OWLReasoner` is the main interface for interacting with OWL reasoners. It provides methods to check satisfiability of classes or ontologies, to compute class and property hierarchies and to check whether axioms are entailed by a KB.

The benefits of porting traditional Semantic Web reasoners like *FaCT++* [20] to mobile platforms should be questioned, as they were designed primarily to run inference services such as classification and consistency check on large ontologies and/or expressive DLs. In ubiquitous contexts, ABox reasoning and non-standard inference services are often more useful, because mobile agents must provide on-the-fly answers to usually smaller problems in moderately expressive KBs [18]. On the other hand, importing a C/C++ library for RDF parsing can be a sensible choice to build an OWL manipulation library or a reasoner. Among the many available tools, the *Redland* [2] suite stands out for functional completeness, standards compliance and code maturity. Other tools like *owlcpp* [9] are less suitable for working in an OWL API port, as they only parse individual RDF triples.

---

[4] GitHub repository: https://github.com/sisinflab-swot/OWL-API-for-iOS.
[5] Apache Jena project: https://jena.apache.org/.

3   Reasoning on iOS Devices: OWL API Porting

The proposed software is a port of the OWL API version 3.2.4. It was implemented in Objective-C –deemed as more mature and stable than Swift– as an *iOS Framework, i.e.*, a library easily used by applications through dynamic linking. The following subsections report on the general architecture and devised performance optimization, respectively.

3.1   Models and Architecture

The OWL API entry point is the `OWLManager` class implementing the `OWLOntologyManager` interface, which allows loading and manipulating a KB. As shown in Fig. 1, the library architecture includes two basic components, the *OWL Model* and the *OWL Parser*. Java *interfaces* were translated to the corresponding Objective-C *protocols*, therefore the Model is interface-wise as the one of the OWL API. The current version does not model the whole OWL 2 language, but a fragment of it exhaustive enough to manage KBs in the $\mathcal{ALEN}$ DL with role hierarchies. In more detail, classes; property restrictions; Boolean class expressions; object properties; declaration, subclass, disjointness, equivalence, domain, range, class assertion and object property assertion axioms are modeled.

The Parser module uses the *Raptor* RDF parser from Redland to deserialize RDF/XML documents (other syntaxes were not considered at this early stage)

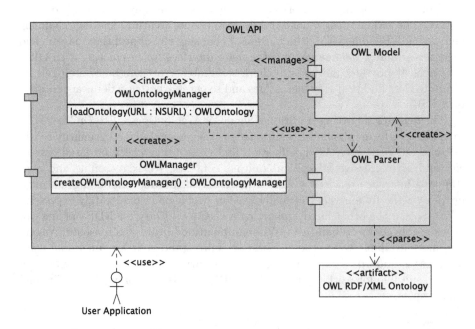

Fig. 1. Main components of the ported library

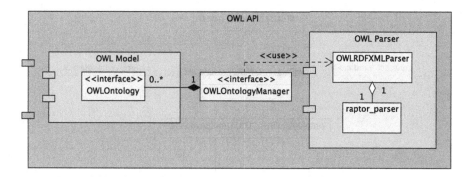

Fig. 2. Detail of the interaction between the Model and Parser modules

into streams of RDF statements. The OWLOntologyManager invokes Raptor through an OWLRDFXMLParser wrapper, which further processes the RDF statement stream in order to create an in-memory representation of the referenced OWL constructs and returns a fully populated OWLOntology object. The interaction between the Model and Parser modules is detailed in Fig. 2.

OWL ontology parsing from RDF triples does not follow the original OWL API approach. A simpler and leaner architecture was adopted, particularly fit for small and medium sized KBs. The implementation of OWLOntology interface is built through the OWLOntologyInternals class, which is populated incrementally during the parsing. It contains data structures such as maps and sets. As pictured in Fig. 3, OWLStatementHandlerMap associates each type of statement to a proper handler, as allowed by the Raptor library. Handlers are implemented as Objective-C *blocks*, which are similar to Java *lambdas* or C *function pointers*. Furthermore, the *builder* pattern was adopted to create instances within the Model component incrementally, because OWL axioms can derive from a variable number of RDF statements.

3.2   Optimization

Optimization effort basically focused on an efficient use of memory, which is the most constrained resource on mobile devices. Execution time was also profiled and optimized wherever possible. In what follows followed optimization directions are outlined.

Architectural optimization. The whole Model component is composed of *immutable* objects. This allows having just one copy of every instance in memory, saving space and time; moreover, it makes the whole component threadsafe. With immutable objects, object hashes can be cached to speed up the very frequent accesses to associative data structures. As a further optimization, if one guarantees that equal objects have the same memory address, the address itself is a perfect hash and equality check becomes just a pointer comparison. In order to make this property true, the library uses the *NSMapTable*

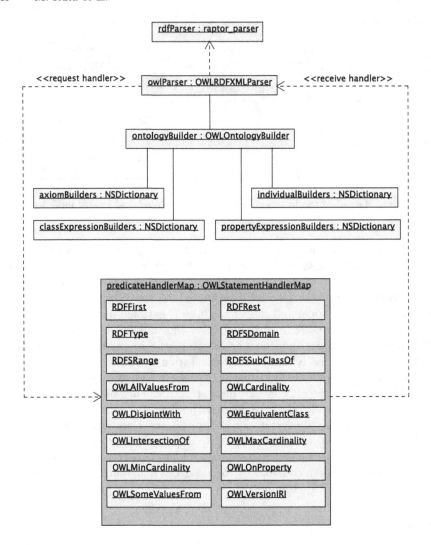

Fig. 3. Main objects of the Parser module

class of the Objective-C Foundation framework as hash table, which supports pointer identity for equality and hashing. NSMapTable was set up to use *weak* references to allow de-allocation of unused objects. This approach, however, is beneficial only in hash tables with low collision rates: this was not found out to be true for all OWL API model classes. Therefore it was adopted just for entities (classes, object properties, individuals) and some axioms considered as performance-critical after profiling tests. These optimizations allowed to roughly halve the measured parsing turnaround times w.r.t. the initial implementation. **Parsing optimization.** During the parsing process, each RDF triple is wrapped in a RDFStatement instance, which is discarded as soon as it is not used anymore.

Furthermore, builders cache the objects they populated, saving both time and memory (in case of similar but not identical instances). Finally, axiom builders are de-allocated in groups: this reduced the observed memory usage peak during parsing by about 30% in preliminary tests.

4   Experiments

The formal correctness and completeness of results provided by the iOS library was evaluated on a set of 34 KBs, obtained from the 2012 OWL Reasoner Evaluation Workshop reference dataset[6] considering all the KBs in the supported $\mathcal{AL}$, $\mathcal{AL}+$ and $\mathcal{ALE}$ DLs. The original Java OWL API 3.2.4 was leveraged as a test oracle. After parsing, the following tests were performed against each KB as significant examples: (i) retrieval of all axioms; (ii) retrieval of all axioms of a given kind; (iii) retrieval of all classes, individuals and properties; (iv) retrieval of all disjoint, equivalent and subclass axioms. The iOS library correctly parsed every KB in the test set, and the returned output of all retrieval tasks proved to be equivalent to the Java OWL API.

Performance evaluation was carried out on a subset of the KBs used for the correctness tests, reported in Table 1. They were selected because they are representative of both traditional and SWoT scenarios, while allowing to sample the performance of the iOS library when working with KBs of varying size. For each KB, three tests were performed: (i) parsing turnaround time; (ii) memory usage peak; (iii) query turnaround time. Each test was repeated five times: for turnaround time tests, the average of all runs was taken. For memory tests, the final result is the average of the last four runs, in order to consider a worst-case scenario due to potential memory leaks. Test devices are listed in Table 2.

Table 1. Knowledge bases used in the performance tests.

Knowledge base	DL	Category	Axioms	Size (kB)
spider_anatomy.owl	$\mathcal{ALE}$	Small	1392	187
brenda.owl	$\mathcal{ALE}$	Medium	14262	1515
mammalian_phenotype.owl	$\mathcal{AL}+$	Large	46081	4289
teleost_taxonomy.owl	$\mathcal{AL}$	Large	195351	21878

Figure 4 shows the results of parsing turnaround time tests: times grow linearly with the size of the parsed ontologies, and small-to-medium ontologies are parsed in about one second or less on devices more than two years old (iPhone 5s). This result is aligned with the performance goals of a mobile reasoner, especially considering that parsing only happens once per usage session, rather than each time a query is submitted to the reasoner.

---

[6] http://www.cs.ox.ac.uk/isg/conferences/ORE2012/.

Table 2. Devices used for performance evaluation.

Device	OS	CPU	Arch.	RAM
Retina MacBook Pro 2014	OS X 10.11.5	Intel Core i7-4870HQ@2.5 GHz	64 bit	16 GB DDR3@1600 MHz
iPhone 6s	iOS 9.0.2	Apple A9@1.8 GHz	64 bit	2 GB LPDDR4
iPhone 5s	iOS 9.3.2	Apple A7@1.3 GHz	64 bit	1 GB LPDDR3
iPhone 5	iOS 9.3.2	Apple A6@1.3 GHz	32 bit	1 GB LPDDR2E

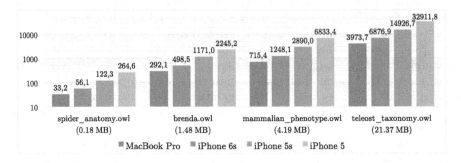

Fig. 4. iOS API parsing turnaround time (ms).

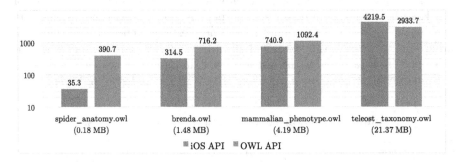

Fig. 5. Comparison of the parsing turnaround time between the iOS API and OWL API (ms).

Figure 5 compares parsing times provided by the iOS API with OWL API on the MacBook Pro testbed. First-run results were considered in this test only, in order to evaluate parsing performance in real usage, since a KB is usually loaded once and queried multiple times. Subsequent runs would provide less realistic results due to in-memory caching. The iOS API shows competitive performance on every test KB, outperforming the OWL API when parsing the small to medium-large ones.

Figure 6 reports on memory usage peak during parsing, which grows linearly with the size of the parsed ontology. Measured values are roughly similar on MacBook Pro, iPhone 6s and iPhone 5s, while they are about 40% lower on

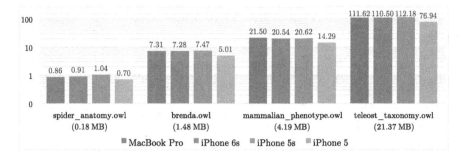

Fig. 6. Memory peak while parsing (MB).

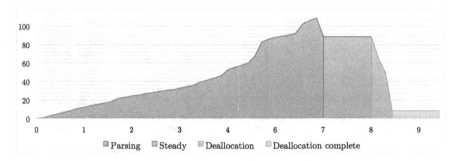

Fig. 7. Memory usage (MB) as a function of time (s).

iPhone 5: this is likely due to it being the only 32-bit device among the four. The results of this test were overall satisfactory, since the required memory is consistent with RAM availability of modern iOS devices.

Figure 7 shows the memory usage trend while parsing and querying the largest KB in the test set (*teleost_taxonomy.owl*) on iPhone 6s. Four phases can be pinpointed: memory usage raises and reaches its peak value during the **parsing** phase; during the **steady** phase the KB is fully loaded and can be queried; memory is released when the KB is **de-allocated**.

Figure 8 shows the turnaround times for the retrieval of all classes in the ontology. This specific query is unrealistic, but it was chosen nonetheless as a stress test for the library. As also seen in the previous tests, times grow linearly with the size of the queried ontology. In order to contextualize the obtained results, query times were compared to OWL API on the MacBook Pro testbed: as reported in Fig. 9, the iOS API outperformed OWL API on every test ontology, confirming its suitability to be used in mobile and pervasive scenarios.

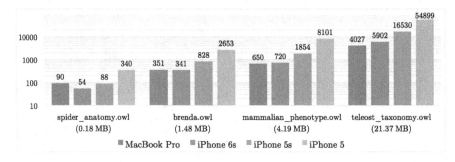

Fig. 8. All classes retrieval query turnaround time (μs).

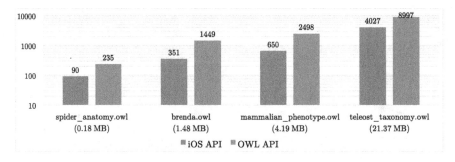

Fig. 9. Comparison of the turnaround time for all classes retrieval query between the iOS API and OWL API (μs).

5   Conclusion and Future Work

The paper presented early results of porting the OWL API to Objective-C, targeting mobile reasoning on the iOS platform. The developed library can run unmodified also on macOS. Early experiments on a small set of ontologies showed correctness of implementation and satisfactory performance in KB parsing and manipulation.

In its current form, the proposed library is ready to support the port of the Mini-ME mobile matchmaking and reasoning engine [18] to iOS, which was the main motivation for the endeavor and is the first planned future work. As a further hope, it will benefit the community as a whole and –possibly with the help of other developers– will grow toward a complete port, aligned with latest OWL API version.

Acknowledgment. The authors acknowledge partial support of Apulia region cluster project PERSON (PERvasive game for perSOnalized treatment of cognitive and functional deficits associated with chronic and Neurodegenerative diseases).

References

1. Apple Inc.: Apple celebrates one billion iPhones. http://www.apple.com/newsroom/2016/07/apple-celebrates-one-billion-iphones.html. Accessed 15 Sep 2016
2. Beckett, D.: The design and implementation of the Redland RDF application framework. Comput. Netw. **39**(5), 577–588 (2002)
3. Bobed, C., Yus, R., Bobillo, F., Mena, E.: Semantic reasoning on mobile devices: do Androids dream of efficient reasoners? Web Semant. Sci. Serv. Agents World Wide Web **35**, 167–183 (2015)
4. Brickley, D., Guha, R.V.: RDF schema 1.1. W3C Recommendation **25**, 2004–2014 (2014). https://www.w3.org/TR/rdf-schema/
5. Giannakis, M., Giannakis, M., Louis, M., Louis, M.: A multi-agent based system with big data processing for enhanced supply chain agility. J. Enterp. Inf. Manage. **29**(5), 706–727 (2016)
6. Hillerbrand, E.: Semantic web and business: reaching a tipping point? In: Workman, M. (ed.) Semantic Web: Implications for Technologies and Business Practices. Springer, Heidelberg (2016)
7. Horridge, M., Bechhofer, S.: The OWL API: a Java API for OWL ontologies. Semant. Web **2**(1), 11–21 (2011)
8. Knublauch, H., Fergerson, R.W., Noy, N.F., Musen, M.A.: The Protégé OWL plugin: an open development environment for semantic web applications. In: McIlraith, S.A., Plexousakis, D., Harmelen, F. (eds.) ISWC 2004. LNCS, vol. 3298, pp. 229–243. Springer, Heidelberg (2004). doi:10.1007/978-3-540-30475-3_17
9. Levin, M.K., Cowell, L.G.: owlcpp: a C++ library for working with OWL ontologies. J. Biomed. Semant. **6**(1), 1 (2015)
10. Liebig, T., Luther, M., Noppens, O., Wessel, M.: Owllink. Semant. Web **2**(1), 23–32 (2011)
11. Matentzoglu, N., Leo, J., Hudhra, V., Sattler, U., Parsia, B.: A survey of current, stand-alone OWL reasoners. In: Informal Proceedings of the 4th International Workshop on OWL Reasoner Evaluation, vol. 1387 (2015)
12. Mohamed, I., Patel, D.: Android vs iOS security: a comparative study. In: 2015 12th International Conference on Information Technology - New Generations (ITNG), pp. 725–730 (2015). doi:10.1109/ITNG.2015.123
13. Noppens, O., Luther, M., Liebig, T., Wagner, M., Paolucci, M.: Ontology-supported preference handling for mobile music selection. In: Proceedings of the Multidisciplinary Workshop on Advances in Preference Handling, Riva del Garda, Italy (2006)
14. Pfisterer, D., Römer, K., Bimschas, D., Kleine, O., Mietz, R., Truong, C., Hasemann, H., Kröller, A., Pagel, M., Hauswirth, M., et al.: SPITFIRE: toward a semantic web of things. IEEE Commun. Magaz. **49**(11), 40–48 (2011)
15. Ploennigs, J., Schumann, A., Lécué, F.: Adapting semantic sensor networks for smart building diagnosis. In: Mika, P., et al. (eds.) ISWC 2014. LNCS, vol. 8797, pp. 308–323. Springer, Heidelberg (2014). doi:10.1007/978-3-319-11915-1_20
16. Schreiber, G., Raimond, Y.: RDF 1.1 Primer. W3C Working Group Note (2014). https://www.w3.org/TR/rdf11-primer/
17. Scioscia, F., Ruta, M.: Building a semantic web of things: issues and perspectives in information compression. In: Semantic Web Information Management (SWIM 2009), Proceedings of the 3rd IEEE International Conference on Semantic Computing (ICSC 2009), pp. 589–594. IEEE Computer Society (2009)

18. Scioscia, F., Ruta, M., Loseto, G., Gramegna, F., Ieva, S., Pinto, A., Di Sciascio, E.: A mobile matchmaker for the ubiquitous semantic web. Int. J. Semant. Web Inf. Syst. (IJSWIS) **10**(4), 77–100 (2014)
19. Steller, L., Krishnaswamy, S.: Pervasive service discovery: mTableaux mobile reasoning. In: International Conference on Semantic Systems (I-Semantics), Graz, Austria (2008)
20. Tsarkov, D., Horrocks, I.: FaCT++ description logic reasoner: system description. In: Furbach, U., Shankar, N. (eds.) IJCAR 2006. LNCS (LNAI), vol. 4130, pp. 292–297. Springer, Heidelberg (2006). doi:10.1007/11814771_26
21. W3C OWL Working Group: OWL 2 Web Ontology Language Document Overview (Second Edition), W3C Recommendation (2012). https://www.w3.org/TR/owl2-overview/
22. Weiß, F., Leimeister, J.M.: Why can't i use my iphone at work?: managing consumerization of IT at a multi-national organization. J. Inf. Technol. Teach. Cases **4**(1), 11–19 (2014). doi:10.1057/jittc.2013.3

Correction to: Use Cases and Suitability Metrics for Unit Ontologies

Markus D. Steinberg, Sirko Schindler, and Jan Martin Keil

Correction to:
Chapter "Use Cases and Suitability Metrics for Unit Ontologies" in: M. Dragoni et al. (Eds.): *OWL: Experiences and Directions – Reasoner Evaluation*, **LNCS 10161,**
https://doi.org/10.1007/978-3-319-54627-8_4

Chapter, "Use Cases and Suitability Metrics for Unit Ontologies" was previously published non-open access. It has now been changed to open access under a CC BY 4.0 license and the copyright holder updated to 'The Author(s)'. The book has also been updated with this change.

The updated original version of this chapter can be found at
https://doi.org/10.1007/978-3-319-54627-8_4

© The Author(s) 2023
M. Dragoni et al. (Eds.): OWLED-ORE 2016, LNCS 10161, p. C1, 2023.
https://doi.org/10.1007/978-3-319-54627-8_12

Author Index

Printed in the United States
by Baker & Taylor Publisher Services